D1053378

Connected Knowledge

Connected Knowledge

Science, Philosophy, and Education

ALAN CROMER

New York · Oxford
OXFORD UNIVERSITY PRESS
1997

Oxford University Press

Oxford   New York

Athen   Auckland   Bangkok   Bogotá   Bombay
Buenos Aires   Calcutta   Cape Town   Dar es Salaam
Delhi   Florence   Hong Kong   Istanbul   Karachi
Kuala Lumpur   Madras   Madrid   Melbourne
Mexico City   Nairobi   Paris   Singapore
Taipei   Tokyo   Toronto

and associated companies in
Berlin   Ibadan

Copyright © 1997 by Alan Cromer

Published by Oxford University Press, Inc.
198 Madison Avenue, New York, New York 10016
Oxford is a registered trademark of Oxford University Press

All rights reserved. No part of this publication may be reproduced,
stored in a retrieval system, or transmitted, in any form or means,
electronic, mechanical, photocopying, recording, or otherwise,
without the prior permission of Oxford University Press.

Library of Congress Cataloging-in-Publication Data
Cromer, Alan H., 1935
Connected knowledge: science, philosophy, and education /
Alan Cromer.
p.   cm.
Includes bibliographical references and index.
ISBN 0-19-510240-1 (cloth)
1. Science—Philosophy. 2. Education—Philosophy. 3. Science—
Study and teaching—Philosophy. 4. Science—Social aspects. I. Title.
Q175.C894 1997
507'. l—dc20   96-14270

9 8 7 6 5 4 3 2 1
Printed in the United State of America
on acid free paper

In memory of my parents
Bert and Helen
who gave me life, love, humor, and intelligence
and of our housekeeper
Nellie Dick Behrendt
who gave me shelter from the storm

Preface

Schools teach, and most people believe, that science is the process of explaining phenomena by making inferences about their causes. In this view, science is a rather natural human activity that's pursued, to some degree or other, by all peoples. But as reasonable as this sounds, it fails to distinguish science from other intellectual activities. As a consequence, science education in the United States fails to develop in students the unique habits of mind that constitute scientific thinking.

In *Uncommon Sense: The Heretical Nature of Science,* I argued that scientific thinking is very different from traditional thinking and that it developed from the particular cultural and historical circumstances of ancient Greece. In this book, I present a discussion of the philosophy of science and its relation to education from the viewpoint of a classroom teacher with forty years of experience teaching graduate and undergraduate physics. I'm a pop philosopher, not a proper Popper one. Since 1990, I have also been engaged in Project SEED (Science Education through Experiments and Education), a program to enhance the knowledge of physical science of middle-school science teachers, and in Project RE-SEED (Retirees Enhancing Science Education through Experiments and Education), a program that trains retired engineers and scientists who volunteer to assist middle-school science teachers in teaching activity-based science. From these involvements, I can view, with nearly Homeric perspective, the educational process from ages ten to seventy.

The simple experiments of middle-school science and introductory college physics exemplify with particular clarity certain issues in the philosophy of science—especially those concerning the meaning of objectivity. By keeping the discussion at this level, the connection between the philosophy of science and the teaching of science is immediately apparent. One sees at once the coequal roles that theory and experience (observation and experiment) play in science. Indeed, science is the knowledge that arises from the connection of theory and and experience.

An explanation is possible only within a theoretical framework and this framework, even for such a simple matter as floating and sinking, is beyond the ability of most children—indeed, of most scientists—to erect from scratch. I understand Archimedes' principle of buoyancy because I've been taught it, but I'm no Archimedes. I have to stand on my tiptoes just to glimpse his immense genius. Yet students are urged to "explain" their haphazard observations of floating and sinking anyway they can. The result, far from empowering them, makes them slaves of their own subjective opinions.

The modern philosophy of science arose from attempts to reconcile the emerging concepts of quantum physics with the positivist doctrines of the German physicist-philosopher Ernst Mach. This work has been largely overshadowed by the common, but erroneous, belief that quantum mechanics introduced subjectivity and uncertainty into science. Although quantum mechanics is hardly a middle-school subject, misunderstandings of it distort popular beliefs about science and how it should be taught. This subject is treated in Chapter 3, which argues that quantum mechanics inserts far more certainty into our understanding of nature than it takes away.

The attacks on the objectivity of science come largely from some social scientists, who claim that the natural sciences are as subjective as the social sciences. But if there can be no objective knowledge about human behavior, how can there be any rational basis for making public policy? On what basis do we decide how and what to teach our children? The problem with the social sciences isn't so much that human behavior is too complex to be amenable to scientific investigation, but that many social scientists work in isolation from the natural sciences. Untethered to reality, the social sciences have splintered into endless specialties and subspecialties. Yet human social patterns aren't much evolved from those of other primates, and an unflinching look at human social organization reveals universal patterns with fairly obvious survival benefits. We are as inherently rebellious as we are loyal, because the latter promotes the survival of the individual while the former promotes the dispersion of the species.

Most educational reformers accept, at some level, the Rousseauian belief in an ideal natural state in which human beings were once freer and less constrained than they are in organized society. This has led to repeated attempts at progressive, naturalistic, and holistic education, all of which inevitably failed. There never was a time when human beings didn't live in hierarchical groups, and reading, writing, mathematics, and science aren't natural. They are each highly complex subjects, with many component parts that must be learned before the whole can be understood. This rather obvious point has been totally abandoned in much of today's educational practices, perhaps as an overreaction to the failure of programmed learning to live up to its own overhyped claims. But the current belief that the whole can be learned without first learning its parts is as pernicious an idea as ever infected education. It's rapidly turning us into a nation of illiterates.

The simplest learning task—a mouse learning a short path through a simple maze—is far more complex than it appears to be. The minimal requirements for learning a maze are found by developing a computer model that learns a maze the way a mouse does. By comparing the predictions of the model with the behavior of real mice, the roles of trial, error, reward, and understanding can be studied. Surprisingly, there's no evidence that a mouse learns anything about a maze, no matter how long it wanders about in one, until it first reaches the reward. That is, during its exploration of the maze, the mouse isn't developing a map of the maze. Nevertheless, once success is achieved, the mouse does rapidly improve its performance in subsequent trials. It achieves reasonably good performance without acquiring any conceptual understanding at all. But it can't, through its own behavior, eliminate all its errors. This is remarkably similar to the behavior of a student in a physics laboratory—or in most other learning situations—who can't achieve error-free performance without intervention from a teacher. This comparison isn't intended to offend either mouse or student, but to argue that there are some universal patterns of learning.

Pure-bred strains of mice show measurable genetic differences in their maze-learning abilities that are mimicked by my computer model. This drops me right into the middle of the debate on genetics, race, and intelligence. Most pertinent to the theme of this book is the question of whether intelligence is, or can be made to be, a scientific concept. I believe that it can be, arguing that its emergence from the vagaries of everyday language to the ranks of a measurable cognitive attribute parallels the emergence of such physical concepts as temperature, genes, and atoms. Furthermore, in spite of the hostility that greeted *The Bell Curve,* there is actually more substantive agreement among the antagonists than their rhetoric indicates. Once extreme claims are eliminated from both sides, there is agreement that human beings do differ in their cognitive abilities. And once this is admitted, the quantification of intelligence, as imperfect as it may be, must be recognized as one of the most important concepts in the social sciences.

Rather than fleeing from the concept of intelligence, educators should adopt a politics of intelligence that tries to balance the needs of people of different abilities. Like most academics, I have had little experience teaching below-average students. So to get such experience, I have done some volunteer teaching in the county prison. Helping grown men in their struggle with fifth-grade reading and mathematics has helped me appreciate just how complex and mazelike this material is. The politics of intelligence would oppose substituting self-expression for drill and practice, because this destroys below-average students.

I am a critic of the various standard movements in the United States, because they are just job-creating boondoggles for the above average. Like orders for troops to take a highly defended mountain ridge, standards are easy to issue and hard to execute. All the hard work of writing curricula is left to the local school districts, after the various standard groups—both state and national—have spent millions of dollars duplicating each others' efforts. The writing of fanciful goals isn't just wasteful of public money; it's also discouraging of district-level efforts to develop educational programs that are focused, structured, and executable.

In prison, I learned what all the standard writers knew all along: we already have a national standard and the curricula to teach it. It's the tests of General Educational Development, or GED tests. The academic skills needed to pass the GED tests are also those needed for almost any kind of education beyond ninth grade, and most inmates don't have them. A politics of intelligence would focus on this vital and attainable standard, even if this means grouping students by ability.

A reader may wonder about my own political persuasion. I wonder myself. Perhaps I can best be described as a realist. I share the liberal's desire to improve the human condition and I share the conservative's belief that human nature, and the laws of physics, limit what's possible. Politically, this book attempts to reconcile desire and belief.

Boston   A.C.
August 1996

Acknowledgments

I, and the reader, owe a heap of gratitude to my brother, Richard B. Cromer, for his word-by-word comments on two drafts of this book. He constantly challenged me to strive for ever greater clarity, with the result that many hopelessly obscure passages were sent to bit heaven.

My late cousin, Richard F. Cromer, first introduced me to psycholinguists in 1969. He also introduced me to the psychologist A. R. Jonckheere of the University of London, who guided and encouraged the work I did at that time on maze learning.

Many of my thoughts on education have developed through my very close association with Christos Zahopoulos, the director of Project SEED. Many of the examples in this book come from Project SEED, which is the product of our collective consciousness. Zahopoulos is also my source of information on the Greek educational system.

The Australian educator and philosopher, Michael Matthews, has had an important influence on my understanding of the philosophical issues relating to empiricism and constructivism. It's because of him that I've had the courage to speak out against the monolithic constructivist movement. And because of his efforts, the movement is beginning to crack a bit.

I have benefited from conversations and arguments with many people in high educational circles—Marilyn Decker, David Driscol, James Fraser, Pendred Noyce, Christopher Randall, Mark Roosevelt, Ethel Schultz, Michael Silevitch, Michael Zapantis. I have also learned an enormous amount about education from conversations with scores of classroom teachers in the SEED program.

A special thanks to Stefan LoBuglio, Orland Ferandes, and the staff of the Education Department of the Suffolk County House of Correction for their generous support of my efforts there and for introducing me to the world of Adult Basic Education.

The chapters on philosophy, quantum mechanics and the history of education have benefited from the expert comments of Professors Michael Matthews, Jorge José, and James Fraser, respectively. David Vernier kindly read the chapter on technology in education

The entire book has been read by my wife, Janet Cromer, and I'm grateful for her comments and suggestions. Her love and support anchor my life, and I love her even when she's critical of my squeezing such large amounts of philosophy out of some very small instances.

Contents

Connected Knowledge

1

Physics, Philosophy, and Education

In Vienna, in the 1920s and 1930s, an informal group of young philosophers and scientists met to discuss critical issues of science that had been raised by the extraordinary discoveries of relativity and quantum mechanics. Through their books and articles there emerged a philosophy, called *logical positivism,* that placed the certainty of scientific knowledge on as firm a philosophical foundation as it's ever likely to have. Today this achievement is dismissed as misguided, due in large part to the remarkable influence of Thomas Kuhn's *The Structure of Scientific Revolutions* (1970). Even scientists who should know better appear to believe that science will forever be an endless succession of revolutionary new theories.

Typically we read in the education literature:

> Students have to learn that . . . new facts of experience can cause the creating of fundamental new theories revising totally the philosophical concepts of science. The theory of relativity and quantum theory are two impressive examples of this century. Hence, human knowledge about science is not definite but preliminary. (Machold, 1992)

There are a number of problems with this. First, students generally don't know the current extent of the "facts of experience," so they can't appreciate the solidity of the foundations of current scientific knowledge. Second, new facts of experience likely to create fundamental new theories in physics will occur at the subatomic level, and will have no effect on existing textbook physics (Schweber, 1993). Thus, while the statement is true in principle, it misleads most students into thinking that introductory physics in the future will be very different from what it is today. Third, "the creating of fundamental new theories" in the human sciences—anthropology and social psychology, in particular—doesn't require new facts of experience, but an assimilation of what is already known in the natural sciences (Freeman, 1992).

Scientific Knowledge

In *Uncommon Sense,* I pointed out the obvious: that much of our current scientific knowledge is both recent and complete (Cromer, 1993). By this I simply mean that we aren't going to discover a new element,[1] anymore than we are going to discover a new continent. There was a time, not so long ago, when we didn't know about these things, but now we not only do, but we know that our knowledge of continents and elements is complete. This to me is one of the important messages of science: we know some important stuff.

The ancient Greeks knew two important facts about the earth: first, that it's round, and second that Europe, Africa, and Asia form a connected land mass, completely surrounded by water. Greek astronomy and geography reached their peaks with Ptolemy in the second century A.D., and in this form it was transmitted to Medieval Europe in the twelfth century, where it remained unchallenged until the transatlantic voyages of Columbus.

The discovery of continents unknown to the ancients opened the entire Greek legacy to review. In particular, Copernicus challenged Ptolemy's earth-centered astronomy. This was the beginning of modern science, since the correct laws of motion could not be found until the entire geocentric cosmography of Aristotle and Ptolemy was overthrown. This change in cosmography was so overwhelming that it's properly called the Scientific Revolution. Kuhn, however, saw it only as the biggest of many revolutions in science. He applied the term "revolution" to much smaller changes in viewpoint, concluding on the basis of a few cases, such as the succession of theories of heat in the eighteenth and nineteenth centuries, that science is destined to be forever changing its view of nature.

It is true that the eighteenth-century phlogistic theory of heat was replaced by successively better theories of chemistry and thermodynamics developed in the nineteenth century, but no one seriously believes that the modern theory of heat, based as it is on the statistical behavior of atoms, will ever be overthrown. Discovery and completion in science is more analogous to the discovery of the new world than it is to political revolution, with its antithesis of counterrevolution. Before the discovery of the Americas, there was no knowledge of other continents, whereas immediately afterwards it became obvious that in time all the continents of the earth would be known.

This analogy is particularly striking in the case of the chemical elements—the material stuff of which the world is made. Medieval Europe inherited from the Greeks the notion that everything is a combination of a small number of simpler or elemental substances. These were usually said to be air, water, earth, and fire, though alchemists in the thirteenth century thought that all metals were the union of mercury and sulfur (Mason, 1962). But it was Lavoisier, in the 1780s, who proposed an operational definition of an element as a substance that couldn't be further decomposed. This is actually quite a sophisticated idea, since it isn't obvious when a substance is heated whether it's reacting or decomposing. Only with the development of techniques for handling gases and precisely weighing matter, could the definition of Lavoisier be put to use.

Lavoisier himself was expert in these techniques and he named twenty-three substances as elements. And for the next hundred years the race was on to find new ones. Between 1790 and 1830, thirty-one new elements were discovered, almost one a year (Mason, 1962). There was no estimate at that time of how many elements there might be, or even if the number was finite. Only with the work of Mendeleev in 1869, based on the sixty elements known by then, was it even acknowledged that the elements followed any order or pattern whatsoever. And only in the 1910s, with the development of atomic theory, did it become clear that the elements form a precisely ordered sequence that depends on the number of protons in the atomic nucleus. This number can be $1, 2, 3, \ldots$, up to 98. Nuclei beyond ninety-eight can be created momentarily, but weighable amounts of the corresponding elements aren't obtained. So there are exactly ninety-eight elements, as "element" was defined by Lavoisier: a weighable substance that can't be further decomposed. We not only know them all, but they and their isotopes have all been studied in great detail. Thus, in the sense that we have complete knowledge of the number of continents, we have complete knowledge of the elements. Furthermore, our knowledge of the properties of the elements is more complete than is our knowledge of the properties of the continents, since the structure of atoms is simpler than the structures of rivers, mountains, and canyons.

Similarly, before the structure of DNA was discovered by James Watson and Francis Crick in 1953, the chemical basis of hereditary was unknown. After the discovery, the structure told us that the genetic blueprint of most organisms was coded by the order of base pairs in their DNA. The human genome is coded by a few billion base pairs, arranged on twenty-three chromosome pairs. Although the number of base pairs is large, it is finite and even manageable, making it likely that the complete structure of the human genome will be known early in the twenty-first century. Presently, the discovery of a new human gene every few weeks gives to biology the same excitement that the discovery of new elements gave to nineteenth-century chemistry.[2]

The deep insights into the nature of matter and living organisms that scientific inquiry have given us also tell us what isn't possible. Although we can make endless combinations of the elements, all materials, at normal temperature and pressure, are limited in density, tensile strength, melting point, and so on by the maximum mass of nuclei and the intrinsic strength of the chemical bond between atoms. Similarly, the base pairs of DNA can form an unlimited array of living organisms, all of which exist within a limited range of temperature because they are all made of carbon, oxygen, and hydrogen.

On the other hand, there may never be a way to know—even in principle— the structure and properties of all possible substances or organisms, just as there is no way to know all the novels that can be written. This is just to say that the ways of combining even a few basic entities can be unimaginably vast, albeit still finite.

Combinations

Let me make a brief technical digression here. George Gamow's *One, Two, Three ... Infinity* (1961) had a strong influence on me as a high school student. In it, he showed that the number of distinct lines of sixty-five characters that can be typed using just fifty keyboard characters (twenty-six letters, ten numbers, and fourteen punctuation marks) is 50^{65}. This is because the character in the first position can be any one of the fifty keyboard characters, so there are fifty possibilities. The character in the second position can be any one of the fifty characters as well, so there are $50 \times 50 = 50^2 = 2,500$ possibilities for the characters in the first two positions. Continuing the argument, there are $50 \times 50 \times 50 = 50^3 = 125,000$ possibilities for the characters in the first three positions, and $50^{65} = 10^{110}$ possibilities for the characters in a full typewritten line of sixty-five characters.

By comparison, the earth is made of fewer than 10^{50} atoms. The entire solar system—the sun and all the planets—has a million (10^6) times as many atoms, or 10^{56}. Continuing the argument, the Milky Way galaxy contains a hundred billion (10^{11})stars, each with about 10^{56} atoms, so the whole galaxy contains $10^{56+11} = 10^{67}$ atoms. And the whole universe has perhaps a hundred billion galaxies, or $10^{67 + 11} = 10^{78}$ atoms. Even if I'm wrong by a factor of ten, or even a hundred, the number of atoms in the universe is only 10^{80}, which is very much smaller than 10^{110}, the number of possible sixty-five character lines that can be written with fifty characters. Thus the number of combinations of even a few items is unthinkably immense.

A computer can easily store billions of numbers; indeed, gigabyte hard-disk drives are commonplace. But no computer could store 10^{110} numbers, even if each memory cell was a single atom and the computer used all the atoms in the universe. It requires exponential notation to discuss such ideas, because ordinary language doesn't have the vocabulary to distinguish a big number, like a trillion (10^{12}) from a super big number, like 10^{110}.

A common notion is that a monkey, typing long enough, could produce a Shakespearean sonnet. Just the opposite is the case. If every monkey on earth typed a line a second since monkeys evolved fifty million years ago, none of the 10^{24} lines produced would be grammatical English, let alone Shakespearian.[3] Thus, although we can know all the characters on a typewriter, and even catalog all the English words in an unabridged dictionary, we can't know all the stories and books that can be written.

In the same sense, we can have complete knowledge of things that we can reasonably count: the continents, the elements, even human genes, because numbers like 7, 98, 100,000, even 3 billion, are manageable given modern computing techniques. But the number of combinations of ninety-eight elements, although still finite, is so large as to be beyond enumeration. Thus, while we have complete and certain knowledge about the atoms of which all matter is composed, we will never know all the ways the atoms can combine to form complex molecules and substances.

Nevertheless, we can say some things with absolute certainty about all substances we'll ever synthesize on earth. For example, no substance will remain solid above a temperature of 5000° C, because there is maximum cohesive force that the electrons can create between atoms. The melting temperature of solids limits the maximum operating temperature of turbines and hence the maximum thrust of rocket engines.

All this is to say that although there may be no limit to the number of exciting new materials that will come from material science, this plethora of wonders should not lead us to the false assumption that anything is possible. What technology does do, is bring us ever closer to the inherent limits of our materials.

Limits of Technology

An important example of this is the efficiency of steam engines. The modern steam engine is the steam turbine, which drives the electrical generators in both nuclear and fossil-fuel power stations. The early eighteenth-century steam engines of Thomas Newcomen had a efficiency of 1 percent, meaning that for every unit of useful work the engine did in lifting water from a mine, it consumed one hundred units of heat. The engine of James Watt, which revolutionized mining and manufacturing, had an efficiency of four percent. Even so, miners were reluctant to replace their Newcomen engines, so Watt and his partner Matthew Boulton resorted to leasing their engine for a payment of half the money saved in fuel costs.

Stimulated by the need for ever-more-efficient engines, scientists in the nineteenth century developed the modern theory of thermodynamics, which gives the precise relations among work, heat, and temperature. Most importantly, it showed how efficiency could be increased by increasing the operating temperature of the engine. Since water boils at 100° C (212° F) at atmospheric pressure, to increase the temperature of the steam further it's necessary to pressurize the system, as in a pressure cooker. The use of pressurized steam started in the early nineteenth century to operate trains and ships, a development strongly opposed by Watt because of the danger of boiler explosions. Indeed, deaths by boiler explosions were common throughout the nineteenth century, and occur occasionally even today.

Modern power stations boil water at pressures of 165 to 218 atmospheres (2400 to 3200 psi), and superheat the resulting steam to a temperatures of 537° C (1000° F). This extremely high-pressure, high-temperature steam drives large turbines, converting heat into work with an efficiency of nearly 40 percent

(Baily, 1977), or ten times Watt's efficiency. But this is the limit; greater efficiencies aren't possible because the highest boiling temperature of water (374° C) occurs at 218 atmospheres. And power plants have been operating near or at this maximum efficiency for most of this century. The importance of this achievement can't be overstated. The world's prosperity is directly related to the cost of fuel and the efficiency of converting the energy in the fuel into useable work. The maximum efficiency was reached ninety years ago and the world's production of fossil fuel may have already peaked (Duncan, 1994). That extraordinary things have been achieved in this century should not blind us to the fact that there are only so many rabbits that a magician can pull from one hat.

This notion of limits, inherent in the achievements of science itself, is hotly disputed by some scientists. In criticizing my views, physicist and science writer Chet Raymo wrote that "in the year 2093 the physics of the year 1993 will seem as partial and tentative as the physics of Aristotle seems to us today" (1993).

The best refutation of this is to look back and see what physicists were teaching 100 years ago. Conveniently, Edwin Hall, in 1897, published a list of sixty-one recommended physics experiments, thirty-five of which were required for admission to Harvard University (Table 1-1) (Moyer, 1976). For anyone who has ever taken a physics laboratory in the last hundred years, the list is embarrassingly familiar. Most of the experiments are still conducted, in one form or another, in a middle school, high school, or college physics laboratory today.[4]

Raymo calls me a pessimist and a know-it-all. I would call myself an optimist and a know-it-all. I take great pride in the knowledge that science has given us. In only a few centuries, it has revealed the discrete, modular, and hierarchical organization of the world. For example, all living organisms are composed of identical and similar cells; all cells are composed of identical and similar protein molecules; all protein molecules are composed of identical and similar amino acids; and so on, down to electrons, protons, and neutrons. It's because of this modularity, and the mathematics of combinations, that the world is knowable, and it's because science has the keys to this knowledge that society values science and science education.

What Raymo and others find upsetting about my view is that it limits the possible. Specifically, it limits hopes of ever communicating with, let alone visiting, reputed intelligences on other stars (Cromer, 1993). More seriously, it limits the hope that there will always be another technological fix for humanity's problems. It just may be that people can get themselves into trouble faster than science can bail them out.

Raymo, like author-scientist Carl Sagan, might be called an optimistic Kuhnian.[5] Optimistic Kuhnians believe there will forever be scientific revolutions that bring us ever more knowledge and power. Kuhn himself didn't believe this, since his theory of revolutions opposes the notion that science is accumulative or progressive (Kuhn, 1970). For him, revolutions weren't necessarily breakthroughs to higher understanding, but the complete replacement of one set of ideas with another.

Table 1-1

Edwin Hall's "Harvard Descriptive List," as lengthened and revised in 1897. The first 25 experiments, which are easier and use simpler equipment, are intended for young children (Moyer, 1976).

MECHANICS AND HYDROSTATICS

1. Weight and unit volume of a substance.
2. Lifting effect of water upon a body entirely immersed in it.
3. Specific gravity of a solid body that will sink in water.
4. Specific gravity of a block of wood by use of a sinker.
5. Weight of water displaced by a floating body.
6. Specific gravity by floatation method.
7. Specific gravity of a liquid: two methods.
8. The straight lever: first class.
9. Centre of gravity and weight of a lever.
10. Levers of the second and third class.
11. Forces exerted at the fulcrum of a lever.
12. Errors of a spring balance.
13. Parallelogram of forces.
14. Friction between solid bodies (on a level).
15. Coefficient of friction (by sliding on incline).

LIGHT

16. Use of Rumford photometer.
17. Images of a plane mirror.
18. Images formed by a convex cylindrical mirror.
19. Images formed by a concave cylindrical mirror.
20. Index of refraction of glass.
21. Index of refraction of water.
22. Focal length of a converging lens.
23. Conjugate foci of a lens.
24. Shape and size of a real image formed by a lens.
25. Virtual image formed by a mirror

MECHANICS

26. Breaking strength of a wire.
27. Comparison of wires in breaking tests.
28. Elasticity: stretching.
29. Elasticity: bending; effects of varying loads.
30. Elasticity: bending; effects of varying dimensions.

31. Elasticity: twisting.
32. Specific gravity of a liquid by a balancing column.
33. Compressibility of air: Boyle's law.
34. Density of air.
35. Four forces at right angles in one plane.
36. Comparison of masses by acceleration test.
37. Action and reaction: elastic collision.
38. Elastic collision continued: inelastic collision.

HEAT

39. Testing a mercury thermometer.
40. Linear expansion of solid.
41. Increase of pressure of a gas heated at constant volume.
42. Increase of volume of a gas heated at constant pressure.
43. Specific heat of a solid.
44. Latent heat of melting.
45. Determination of the dew-point.
46. Latent heat of vaporization.

SOUND

47. Velocity of sound in open air.
48. Wave-length of sound.
49. Number of vibrations of a tuning fork.

ELECTRICITY AND MAGNETISM

50. Lines of force near a bar magnet.
51. Study of a single-fluid galvanic cell.
52. Study of a two-fluid galvanic cell.
53. Lines of force about a galvanoscope.
54. Resistance of wires by substitution: various lengths.
55. Resistance of wires by substitution: cross-section and [parallel connection].
56. Resistance by Wheatstone's bridge: specific resistance of copper.
57. Temperature coefficient of resistance in copper.
58. Battery resistance.
59. Putting together the parts of a telegraph key and sounder.
60. Putting together the parts of a small motor.
61. Putting together the parts of a small dynamo.

Constructivism

Closer to Kuhn's own view, and much more important to science education, are the pessimistic Kuhnians. These are people who deny the possibility of objective knowledge altogether.

> Indeed, [we] constructivists do not consider knowledge to be an objective representation of an observer-independent world. . . . Constructivists . . . do not accept the idea of truth as correspondence with reality. Modern science does not give us truth; it offers a way for us to interpret events of nature and to cope with the world. (Yager, 1991)

Even the National Research Council of the National Academy of Sciences, in its first printing of a document discussing the intellectual foundations of its science education standards stated that the standards would reflect the "postmodernist view of science" that "questions the objectivity of observations and the truth of scientific knowledge." This was said to be the opposite of logical positivism, which it disparaged as being "characterized by arguments for the objectivity of scientific observations and the truth of scientific knowledge" (National Research Council, 1992; National Science Teachers Association, 1993). After protest from the scientific community, this statement was removed from later printings of the document, and the vaunted *National Science Education Standards,* published in 1996 had a more reasonable statement of the nature of science:

> Because all scientific ideas depend on experimental and observational confirmation, all knowledge is, in principle, subject to change as new evidence becomes available. The core ideas of science such as the conservation of energy or the laws of motion have been subjected to a wide variety of confirmations and are therefore unlikely to change in the areas in which they have been tested. (National Research Council, 1996)

This moderation of philosophy may mark a turning point in the battle between constructivists and realists over control of science education. Still, *National Science Education Standards* is long on constructivist jargon—inquiry methods, open-ended and real problems, reflection, construction of understanding—and short on the specificity of Hall's list (Table 1-1), which established the laboratory-based approach to physics teaching that's practiced to this day (Moyer, 1976).[6]

Constructivism is a postmodern antiscience philosophy that is based on Piaget's work on how children construct concepts and conceptual relations and on the philosophy of two early eighteenth-century opponents of the Scientific Revolution, Giambattista Vico and George Berkeley (Matthews, 1993). It's a form of subjective empiricism that puts its emphasis on the thoughts of the knower and views the search for truth as an illusion. "Knowledge can never be considered true in the conventional sense (e.g. correspond to an observer-independent reality) because it is made by a knower who does not have access to such a reality. . . . From [the constructivist's] perspective, Truths are replaced by viable models—and viability is always relative to a chosen goal." Constructivism redefines knowledge to be whatever individuals, "given the range of present experience within their tradition of thought and language, consider *viable*" (Glaserfeld, 1992).

Such an ideology would be of no interest to scientists and science educators were it not, in effect, the official ideology of the reform movement in the United States and elsewhere. New Zealand is so committed to nonobjectivity in science teaching that the lecture-demonstration tables have been removed from all the science classrooms in the country (Matthews, 1995). This is to prevent teachers from claiming to know more than their students, thus unduly influencing how the students' construct their own knowledge. Constructivism is deeply embedded in many educational institutions in the United States, is supported by the National Science Foundation, and was for a time the official ideology of the science-education reform effort in Massachusetts (Massachusetts, 1994). But when push comes to shove, no one knows how students are to construct their own theories of atoms and electrons, of stars and galaxies, of DNA and genetics. Educators are beginning to recognize the limitations of constructivist ideology as they begin to address the problem of implementing state and national content standards.

This openly antiscientific ideology is still fashionable, however, because it's in step with the many postmodern doctrines that are endemic in academia today (Gross and Levitt, 1994). Constructivism fully supports the view that establishment science is the particular construction of white males because it can argue that what is viable for white males at some historical period may not be viable for other human beings at some other time.

But more important, by devaluing scientific knowledge—bringing it down, so it speak, to the level of everyday knowledge—constructivist educators with no knowledge of science have increased their own power in science education relative to educators with scientific knowledge. In the United States, pre-high-school science education, such as it is, is controlled by professional science educators, trained in schools of education which have been notorious for a hundred years for their low academic standards. Rare is the science educator who knows even the science expected of an eighth grader. It's this group which has enthusiastically endorsed constructivism because it allows them to speak only about process (whatever that is) rather than content (of which they are ignorant). And it's this group that writes the frameworks, standards, and textbooks for elementary and middle schools.

This harsh judgment is borne out by the incredible number of errors, misconceptions, and undefined terms that occur in the most recent spate of middle school textbooks. Mario Iona (1994) has published two pages of errors that he found in *SCIENCE Interactions* (Aldridge et al., 1993), commenting that with eleven authors and countless consultants and reviewers they should have done better. In addition to Iona's list, I have a list of my own, including the erroneous implication that buoyancy depends on how far a totally submerged object is below the surface of the water.

The subject of buoyancy is even more mangled in *Middle School Science & Technology,* a three-volume science series for sixth, seventh, and eighth grades that was written and developed by BSCS with funding from the National Science Foundation (BSCS, 1994a, 1994b, 1995c). In the sixth-grade book, *Investigating Patterns of Change,* we are told: "When objects float, it is because the objects push down with less force than whatever is pushing them up" (BSCS, 1994a). Besides its confusing mixture of the singular and the plural, this sentence exemplifies many of the pedagogic and scientific errors that run through the series. First, the term *force* is used here for the first time. It's never explained or defined, and doesn't even appear in the index. The book provides no basis for understanding what force is, let alone what it means for one force to be less than another. Second, the force with which an object pushes down is not a force on the object. It's the force of gravity on the object—its weight—that should be used. Students often confuse the weight of an object with the force with which the object pushes down, but textbook writers are expected to know better. Third, for any object to remain at rest, the upward and downward forces on it must be equal. If the downward force on a floating object were less than the upward force on it, the object would jump out of the water.[7] Fourth, this erroneous and confusing sentence about objects floating in water is the explanation, in the very next paragraph, of the circulation of upper atmospheric winds from the equator to the poles. This is old-fashioned stuff-it-down-their-throats teaching at its worst. It's a national disgrace that a textbook, developed with federal funding, can't get either the pedagogy or the science right.[8]

Constructivist Education

Constructivists believe that each child can learn the scientific process in a rather straightforward manner by observing patterns and making predictions (BSCS, 1994a). This may sound reasonable, but it's essentially Aristotelian science and is contrary to everything we have learned about science since the Scientific Revolution. I had an opportunity to observe this in June, 1995, when a constructivist was invited to facilitate some inquiry-based lessons on density and buoyancy to twenty-four middle-school science teachers who were participating in a two-week SEED (Science Education through Experiments and Education) program.[9]

He began by giving the teachers a number of equal-volume cylinders of different materials and asking them to determine which sank and which floated. He then had them weigh the cylinders and an equal volume of water, and they found that the sinkers were all heavier than the water and the floaters were all lighter. The obvious conclusion is that heavy objects sink and light objects float. But it's the density, not the weight, that determines whether an object sinks or floats. It's impossible to arrive at this elusive conclusion by investigations of this nature.

The word "density" drifted into the discussion and the formula $d = m/V$ was written on the board. But the abstract concept of density is difficult enough to require its own independent development. It doesn't follow from looking at the weights of different materials of the same volume, and introducing it arbitrarily into a lesson on buoyancy creates mind-numbing confusion.

To clarify the situation, the constructivist had the teachers measure the "floating" force on spherical fishing floats of different sizes. This was done with a spring scale that was attached to a float by a string that passed around the hook of a suction cup that was pressed to the bottom of a container of water. By pulling up on the string with the scale, the attached float was pulled under the water. Because of friction between the string and the hook, the spring-scale reading depended on the angle of the spring, which confused the situation somewhat. Nevertheless, the teachers generally found that the force increased with the size of the float. A discussion then ensued to see what conclusion could be drawn from these results.

Some teachers thought that the floating force depended on the total surface area of a sphere, which was consistent with their results. So was a dependence on weight or volume. Further confusion arose when the teachers floated pieces of clay shaped as boats. What had this to do with density? Was surface tension involved? No explanation was accepted or rejected by the facilitator, who seemed to feel that his role was to insure that the discussion became ever more confusing and divergent. When he was asked how teachers were to resolve all this confusion in their classrooms, he suggested that they have their students do further investigations.

Yet further studies of this kind—perhaps plotting the floating force against the weight of the floats, or perhaps measuring the floating force on objects of different shapes—would only compound the confusion. Unguided by theory, it's virtually impossible to understand buoyancy by random investigations of this sort. Why? First, the floating force that the teachers measured isn't the buoyant force, which Archimedes' principle says equals the weight of the displaced liquid, but it's the buoyant force minus the weight of the object, a more complex entity. Second, schools haven't the time or equipment, teachers haven't the knowledge or experience, and students haven't the interest or patience to fruitfully continue such investigations. Third, each new investigation brings in its own variables and inaccuracies, so the more one investigates, the more confusing matters become.

Clarification comes only when there is an idea or principle general enough to explain a wide range of phenomena. Archimedes' principle has this sort of generality. Archimedes (287–212 B.C.) himself was perhaps the greatest intellect of antiquity, and ordinary mortals are lucky if they can understand his principle, let alone discover it. Although he lived only a hundred years after Aristotle and 1,800 years before Galileo, he was a charter member of the Scientific Revolution. In his book *Floating Bodies I,* Archimedes derived the principle of

buoyancy from a single postulate of hydrostatics (Archimedes, 212 B.C./1897).
A Latin translation of this, and some other of his works, was published in Venice
in 1543, the year Copernicus died and twenty-one years before Galileo was
born. Archimedes' work had a great influence on Galileo, who also developed
physical principles from mathematical idealization, using experiments for purposes of demonstration and verification—not for discovery.

Unfortunately, many science educators are genuinely ignorant of the theoretical structure of science. Many also are ignorant of the standard experimental techniques that, for good and proper reasons, have been used for generations
to teach basic scientific principles. Experiments two through seven in Edwin
Hall's "Harvard Descriptive List" of 1897 (Table 1-1) concern density and
buoyancy. The first two experiments in this sequence involve objects that sink,
because the buoyant force on an object that sinks is much easier to investigate
than the buoyant force on an object that floats. By suspending a rock, a piece
of aluminum, or a lump of clay from a spring scale and lowering it into a container of water, the decrease in weight can be measured and—using an over-
flow container made from a milk carton—compared to the weight of the displaced water. This is the approach that the SEED teacher-enhancement
program takes, and most likely it's the approach that Archimedes himself took,
since he was interested in determining whether the king's gold crown had been
adulterated with silver.

The day after the constructivist gave his three-hour lesson, the same teachers participated in the regularly scheduled SEED workshop on the same subjects. This workshop consisted of a ninety-minute demonstration-discussion
period conducted by Christos Zahopoulos, several hours of laboratory work,
and an hour of final discussion. Thus on consecutive days, twenty-four teachers had extensive exposure to two very different approaches to the same subjects. Five days later they participated in confidential focus groups conducted
by Dr. Paula Leventman, the SEED evaluator. Half the teachers were in one
ninety-minute focus group and half were in another. The full three hours of
discussion was taped and later transcribed, omitting the names of the teachers.
The primary purpose of these focus groups was to evaluate the SEED program
as a whole, but there was also much discussion of the two density and buoyancy workshops.

The evaluation shows clearly that the teachers were confused and angered
by the constructivist's workshop. One teacher, who had recently attended a
week-long workshop on constructivism, became an immediate apostate after
being subjected to three hours of inconclusive inquiry. Here are some teacher
quotes from the evaluator's report (Leventman, 1995):

> The difference between the [constructivist's] demonstration on density and
> buoyancy and Christos's was like night and day. There was very, very little
> content—maybe 5 percent content.

> I don't think there was any content. It has made me reevaluate my own use
> of inquiry-based teaching. I am going to change to much more content with
> a little constructivism thrown in.

I walked away from that class thinking this is great, but what did I learn.

Me too. At first I thought it was great fun but then I realized that I didn't know what we were doing. . . . I walked away being more confused about density and buoyancy than when I started. . . . The next day when Christos did it I felt so much better.

It's interesting that some of the teachers weren't able to adequately judge the inquiry-based approach until they had experienced a reasonable alternative. Although this can by no means be considered a scientific study, it's as good a comparison and evaluation as one ever gets in educational research.[10]

SEED
(Science Education through Experiments and Demonstrations)

The guiding principle of the SEED approach is that the concepts in science build upon one another in a systematic way. We know that density can't be understood before there is understanding of mass, volume, and ratio. And we know that volume can't be understood before area, or area understood before length. So SEED starts with the simplest concepts and builds up in complexity through a sequence of concrete activities. It assumes that many of the teachers in the program are, like their students, largely concrete thinkers. Progress toward more abstract thinking is promoted through investigative activities and the use of simple mathematical techniques (ratios, averages, graphs) to analyze quantitative data. As logical as this sounds, and as effective as it proves to be in practice, it isn't the way curricula are developed in the United States.

The NSF-funded textbook series *Middle School Science & Technology* (BSCS, 1994) is deliberately written on a need-to-know basis; that is, concepts are introduced as required by the context. This too seems reasonable, since its purpose is to connect ideas together in a meaningful way. The problem is that concepts aren't factoids that can be dealt with in a sentence or too. Any reasonable understanding of buoyancy requires prior understanding of weight, volume, density, and forces in equilibrium. Each of these prior concepts requires weeks to develop, and bringing them all together is a major challenge for most middle-school teachers and students. In SEED, the teachers do enough density and buoyancy experiments (measuring the weight of the displaced liquid, measuring the density of solids, liquids, and air, building a column of liquids of different densities, building a hydrometer, and so on) to keep their students busy for weeks. To use a brief and incorrect "explanation" of buoyancy to "explain" the circulation of the atmosphere, as the sixth-grade BSCS textbook does, doesn't provide a contextual development of buoyancy, but only piles meaninglessness upon meaninglessness (BSCS, 1994a).

SEED begins with the basic concepts, or underpinnings (Arons, 1990), that are the foundations of physical science: length, mass, time, area, and volume (Cromer, Zahopoulos, and Silevitch, 1994). From there, it goes into force, density (the ratio of mass to volume), pressure (the ratio of force to area), simple machines, motion, the earth as a planet, elements and compounds, sound, optics, temperature and heat, electricity, and electromagnetism. The teachers participate in many investigations that illustrate and develop the principles of the course (Cromer and Zahopoulos, 1993).

Although SEED is conceptual rather than historical in design, the orderly development of the concepts and principles of physical science naturally follows a roughly historical sequence. It is perhaps not realized how much of middle-school physical science was discovered by Archimedes: area of a circle, volume of a sphere, principle of buoyancy, center of gravity, and the principle of simple machines. And many of the SEED experiments and demonstrations can be found in Galileo (1638/1914), such as the scaling of area and volume, the period of a pendulum, measuring the density of air, and the sagging of a stretched rope weighted in the middle.

Used cautiously, history can be a guide to the teaching of science. During the middle-school years, when student should be developing more abstract modes of thinking, the pre-Newtonian science of Archimedes and Galileo is most appropriate. This is full-blown science in the sense that it uses idealizations and mathematics to describe an idealized world, but it doesn't have the formidable conceptual and mathematical difficulties of Newtonian and post-Newtonian physics. I use the terms "pre-Newtonian," "Newtonian," and "post-Newtonian science" to refer not only to specific historical periods, but, by generalization, to the science from any period that requires a corresponding level of abstract reasoning for full understanding. In this sense, SEED is a pre-Newtonian program. Most of its topics were known to Galileo, and those subjects that came later—Newton's laws of motion and electromagnetism—are developed at the pre-Newtonian level of abstraction. Other work in teacher enhancement suggests that given enough time and support, this level can be reached by 85 percent of middle-school teachers (Aarons, 1995). Under optimal conditions, then, about half of the middle-school students nationwide could be brought to this level. A full study of Newtonian mechanics must be deferred to high school or college, for students who have acquired the necessary mathematical background.

Aristotelianism and Elementary Science Education

Aristotelian science explains the unfamiliar in terms of the familiar (Matthews, 1994). It's thus pure Aristotelianism to say that "scientists construct explanations using evidence and inferences" (BSCS, 1994c), that is, that they work backwards, or inductively, from evidence to some unknown or unfamiliar "explanation." Since the Scientific Revolution, scientists have worked forward, or deductively, from the unfamiliar (the law of universal gravitation, for instance) to the familiar (the motion of the planets). The constructivist belief in the nonobjective nature of science stems from its Aristotelian theory of knowledge. As the constructivist's lesson on density and buoyancy so clearly demonstrated, there can be many conflicting "explanations" of the "evidence." Such an approach produces conflict, not resolution.

Nevertheless, Aristotelian science is an early stage of science and it does work, more or less, for phenomena in which the unknowns are generalizations that can be easily inferred from the evidence. For example, the generalization that plants and animals reproduce after their own kind may be inferred from a child's experience with kittens, pups, and gardens. Of course, this is no proof that flies aren't spontaneously generated by garbage, and Aristotle believed that they were (Nordenskiöld, 1928). But since students don't yet have the reasoning skills needed for pre-Newtonian science, elementary science education is necessarily Aristotelian. Unfortunately, science educators don't understand this, or that a mode of instruction appropriate for one age group may not be appropriate for another.

Let me illustrate this with a second-grade activity I observed in a bilingual class in San Francisco in 1993. The children, mostly Chinese-American, were each given one half of a table-tennis ball and instructed to fill it with clay and level it off. They were then shown that if a stick is stuck upright in the clay, the arrangement will, if toppled over, swing upright when released (Fig. 1-1). Their task was to test a variety of objects to see which, when stuck in the clay and toppled, would right themselves, and which wouldn't.

The teacher did an excellent job of collecting each object as it was tested and pasting it on one of two charts, depending on the outcome of the test. She soon had one chart filled with small objects and the other with large objects. The lesson was going splendidly, until two students got contrary results with the same object. A bit flustered with several visiting scientists in her classroom, the teacher asked one of them to "explain" the situation to the children.

While one of the visitors engrossed the second graders with a lecture on the relationship between stability and the height of the center of gravity (complete with a vector diagram), I walked around the classroom comparing several of the

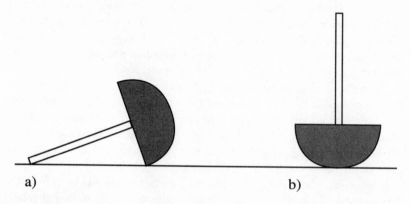

a)   b)

Figure 1-1 A stick inserted into a clay-filled cup made by cutting a table-tennis ball in half. When the cup is toppled over and released, it returns to its upright position.

cut table-tennis balls with one another. They weren't identical; the balls hadn't been cut exactly in half. This alone could explain the contrary results of the students.[11]

This lesson was being shown to a group of scientists as a model of the type of inquiry-based learning being advocated by reformers, although some might find that the teacher controlled the activity too closely for there to be much inquiry going on. All it really showed is that even an experiment designed explicitly for second-graders can have a significant uncontrolled variable (the size of the cut table-tennis ball). The activity is not without scientific interest, but second graders are much too young to control variables, a conceptual skill more appropriate for middle school. By ninth grade, the experiment would be an excellent way to teach about controlling variables, and in a high-school physics course it could be the subject of a general investigation into the many factors affecting the stability of a system. But it has nothing to offer second graders, and could, in fact, be quite detrimental.

For example, confronted with two students getting opposite results with their clay-cup experiment, a teacher might notice that one of the cups was less than half of a table-tennis ball. "See," he says triumphantly, "this is a bad cup. Throw it out and take another." This, of course, is nothing more than scapegoating, a very destructive form of traditional thinking. There are no good and bad cups, only larger and smaller ones. The difference between traditional and scientific thought is the difference between evaluative thinking (good and bad) and objective thinking (larger and smaller).[12]

Language and Understanding

The clay-cup experiment also illustrates, in a very obvious way, the inadequacy of language for discussing science. In English, the word "half" means both the fraction ½ and "one of two, more or less, equal portions of a thing." There is no single word in English that carries the second meaning without the first, or for that thing you get when you try to cut a table-tennis ball in half, but don't succeed. In writing, I could refer to "half" a table-tennis ball, using quotation marks to mean "more or less." In the preceding paragraph, I used "cup," which, if it confused some readers, will have made my point.

Once the concept is understood, finding a suitable means for expressing it isn't a major problem. The major problem is recognizing the concept before one has the word; or, even more confusingly, separating the concept from a word that is also associated with something related. The teacher was given a box of table-tennis balls that had been cut in "half." But since "half" sounds a lot like half, the difference was not recognized.

Indeed, it's commonly believed that the meaning of a word can be found in the word itself. Some teachers like to begin a unit on work by writing the word "work" on the blackboard, drawing a circle around it, and then attaching to the circle other words that their students suggest are related to work. This exercise gets the students involved and the class will rapidly generate a dozen or so associated words: "effort," "job," "force," "lever," "school," and so on. From this, the teacher leads a discussion on what work *really is,* as though the word "work" really is anything at all.

A better approach is to precede any discussion of work with weeks of demonstrations and experiments involving simple machines: levers, pulleys, inclined planes, and so on. Students will find that with a lever, a small force applied far from the fulcrum can lift a heavy weight that's close to the fulcrum. (This led Archimedes, so Plutarch tells us, to remark that if he had a long enough rod, and a place to stand, he could move the earth.) After experimenting with pulley systems and an inclined plane (Fig. 1-2), students are ready to understand that in every simple machine a small input force moving through a large distance causes a large output force to move through a small distance. "Work" is the name given to the number obtained by multiplying a force by the distance through which it moved. For every simple machine, the work done by the input force is equal to, or greater, than the work done by the output force. Only the brightest students may understand this last generality, but most students can, *given enough time,* appreciate that force can be increased indefinitely, at the sacrifice of distance. (Archimedes didn't say how far he intended to move the earth.)

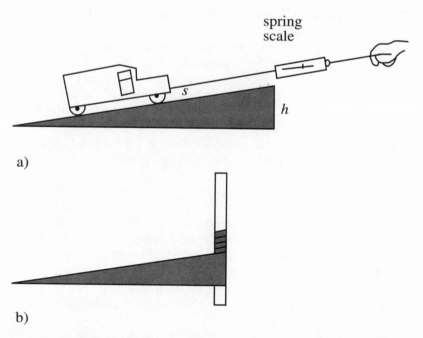

Figure 1-2 a) With a spring scale, a student can measure the smallest force required to pull a toy car up an inclined plane (the input force) and compare it to the weight of the car (the output force). The input force is much less than the output force because it moves along the slope s of the incline as the weight of the car is lifted the vertical distance h. b) By wrapping a paper triangle around a rod, a teacher can demonstrate that a screw is just an inclined plane bent into a spiral.

The word games teachers play at the beginning of a unit aren't as effective as a demonstration in evoking student interest, and they commit the deadly error of confusing the name we give something with the thing itself. This error is common in the teaching of science, and universal in the teaching of everything else. Academics argue endlessly over the meaning of "constructivism," as though the word has an objective meaning that can be discovered by argumentation.

Language appears to have evolved separately from intellect (Chapter 6), which may explain its seductive power. A few patriotic slogans, and millions of people march off to war or to vote for independence from Canada. In rational discourse, words don't stand on their own—they must relate to something real. The purpose of the physics laboratory is to show students the reality behind the words and symbols that are in their textbook. The experience gives them a better understanding of abstract concepts such as work and force, while inculcating a healthy skepticism toward empty abstractions such as national identity and traditional values.

Empiricism and Logical Positivism

Empiricism is the philosophical principle that knowledge must be based on the evidence of our senses, rather than on the pronouncements of priests and prophets. At its extreme, pure empiricism denies the existence of any reality beyond the senses. The positivists Auguste Comte and Ernst Mach were pure empiricists because of their profound distaste and fear of metaphysics, whereas many constructivists are pure empiricist because of their ignorance of the scientific process. Mach (1838–1916) is a central figure in the development of our current understanding of science. He waged a ceaseless battle against the metaphysical fog that then dominated German philosophical thinking and he stimulated a whole band of younger scientists to find a better foundation for science.

But Mach, in his enthusiasm to eliminate all such metaphysical constructs, also attacked the concept of the atom. For Mach, the atom was just another unnecessary pseudoexplanatory notion that served no real purpose. The young scientists and philosophers who met in Vienna in the 1920s and 1930s to discuss these matters—Philip P. Frank, Hans Reichenbach, Rudolph Carnap, Moritz Schlick—strongly supported Mach's attack on empty metaphysics, but they wanted to find a place for atomic theory, quantum mechanics, and relativity which were daily generating new insights and experimental triumphs. They wanted Mach and atoms, too.

The philosophy of science that emerged from the Vienna Circle is called *logical positivism*. *Positivism* is another term for empiricism, and the qualifier *logical* softens the pure empiricism of Mach by allowing for the theoretical structures of modern physics. For the positivist Mach, the purpose of science was to find the correlations between the events we perceive through our senses. To this end, scientific principles might be useful tools of the trade, but he refused to concede that they were true, or approximately true, descriptions of an extrasensory world. The logical positivists, influenced by the developments of relativity and quantum mechanics which had overthrown some of the basic notions of space, time, and causality assumed by Newton, wanted to develop a foundation for science that would never be overthrown by future discoveries. This meant grounding science on experience and eliminating untestable concepts, as Mach had taught them. But unlike Mach, the logical positivists understood the unavoidable role that extrasensory theoretical entities play in connecting different observations. And as some extrasensory entities, such as atoms and genes, became ever more visible, they were less inclined to deny their reality. Indeed, the reality of most entities, from the internal temperature of a star to the age of the earth, is linked to observations only through a long chain of deductive reasoning.

The most basic principle of physics—that a mass will move forever in a straight line at constant speed if no force is acting on it—is hardly plausible, let alone susceptible to empirical verification. Only when Galileo dared to imagine an ideal environment, free from gravity and friction, was he able to visualize this extrasensory truth. Such an ideal environment doesn't exist anywhere in the universe, which may explain why, to this day, it's difficult to present a convincing demonstration of the law of inertia to a physics class.

It's the notion of a theoretical framework that's missing from so much of recent "reform" in science education. In the clay-cup experiment, the second-grade teacher was leading her students to the conclusion that the system was stable (popped back up) when small objects were stuck in the clay, and unstable with large objects. But this conclusion is false. In fact, nothing meaningful can be concluded from such a hodgepodge experiment, and the implication that it can be only reinforces unscientific thinking. Even a properly controlled experiment that used sticks of equal diameter and unequal length would not advance a student's understanding of stability, because such understanding requires a theoretical framework well beyond second graders and second-grade teachers.

Science and Postmodernism

Science and postmodernism fundamentally disagree about the possibility of obtaining objective knowledge. Postmodern constructivists are pure empiricists who deny the possibility of human beings obtaining objective knowledge about an observer-independent reality, whereas science assumes the existence of an objectively knowable external world. All postmodern philosophies, including constructivism, deny the existence of fixed rules. whereas science places human activity within a natural world with immutable rules that are independent of human desires. This disagreement has deep political implications, because all radical movements, whether left or right, assume that there are no rules that can't be broken.

The denial of objectivity can lead to catastrophe, because its logical consequence is chaos or tyranny (Matthews, 1993). In various conversations with his intimates, the professed new-age prophet Adolf Hitler clearly articulated his own scorn for reason and objectivity (Holton, 1996):

> We are now at the end of the Age of Reason. . . .We are at the outset of a tremendous revolution in moral ideas and in men's spiritual orientation. . . . A new age of the magical interpretation of the world is coming, of interpretation in terms of the will and not of the intelligence. . . . There is no truth, either in the moral or in the scientific sense. . . . The idea of free and unfettered science . . . is absurd. . . . Science is a social phenomenon, and like every other social phenomenon is limited by the benefit or injury it confers on the community. . . . The slogan of objective science has been coined by the professorate simply in order to escape from the very necessary supervision by the power of the State. . . . It follows necessarily that there can only be the science of a particular type of humanity and of a particular age. (Rauschning, 1940)

The similarity of these thoughts with those of the postmodern constructivists is chilling; even more chilling is the degree that this thinking has penetrated our educational system. Freedom is in jeopardy when a society ceases to believe in an objective reality, for then all that is left are the egocentric beliefs of its individuals. In such a climate, cults are formed and demagogues arise eager to recruit the gullible and the unthinking.

The aim of democratic education should be the development of objective thinking and productive skepticism. The common bases for these are a decent respect for the opinions of others and a desire to resolve disagreements by further research aimed at reaching informed consensus. Indeed, science has been defined as the search for consensus (Ziman, 1968) in contrast to (say) law, where disputes are resolved by judge and jury, or theology, where disagreement leads to schism and inquisition.

But what exactly is objectivity? Can it be defined with enough specificity to guide us in developing effective ways of teaching it? In the next chapter, we give an answer that has surprisingly wide acceptance in the scientific community.

2

Theory and Experience

A pencil balanced on its tip is a popular demonstration with children and adults of all ages (excluding ages thirteen through sixteen, of course). It's especially fun to watch the delight on a four-year-old's face when the pencil actually balances on her finger. The demonstration is easy to prepare: one end of a piece of wire is wrapped around the top of the pencil and the other end is wrapped around a fishing weight or similar object. The wire is bent, as shown in Fig. 2-1, so that the weight is below the pencil's tip. When adjusted properly, the pencil balances on its tip. Why?

Someone with a physics background might answer that the weight has lowered the center of gravity of the system, and thus and this and so on. But this isn't the sort of explanation I'm interested in right now. I want an explanation suitable for a four-year-old. Such an explanation, far from being oversimplified,

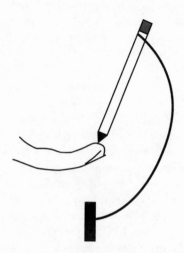

Figure 2-1 A pencil balanced on its tip.

will be exceedingly deep, since it will go beyond the words to the phenomenon itself.

The question of what we mean by an explanation has been discussed by philosophers for centuries. In idealistic philosophy, most notably associated with Emmanuel Kant (1724-1804), explanations are in terms of *a priori* concepts, absolute truths that are known to us by pure thought. The postulates of Euclidian geometry were thought to be examples of such truths, until the development of non-Euclidean geometries by Nikolai Lobachevski and Georg Friedrich Riemann. In Kant's "foggy metaphysics," even the laws of Newtonian mechanics could be derived by pure reason (Franck, 1949).

This wasn't the belief in Newton's time, nor is it in ours. Newton's critics deplored the unsound and unreasonable nature of his laws, especially the law of universal gravitation. They felt it was unscientific to introduce a mystical invisible force that acted across limitless empty space, while his admirers lauded him for discovering a law so contrary to our senses (thus proving that God exists). Newton's great work, the *Principia* was written in Latin and first published in 1686. For the preface to the second edition, published in 1713, Roger Cotes wrote:

> All sound and true philosophy is founded on the appearances of things; and if these phenomena inevitably draw us, against our wills, to such principles as most clearly manifest to us the most excellent counsel and supreme dominion of the All-wise and Almighty Being, they are not therefore to be laid aside because some men perhaps dislike them. (Newton, 1713/1947)

But by Kant's time, Newton's laws had been elevated to the status of self-evident truths by philosophers who never had to teach them to disbelieving freshmen. The development of relativity and quantum mechanics in the first two decades of this century caused the whole idealistic philosophy of explanation to be reconsidered by the logical positivists, who took their lead from Ernst Mach. The aim of this effort was to develop a theory of explanation that wouldn't have to be revised with every new discovery in science.

Experience

All scientifically inclined philosophers agree that science is based on experience, but each uses somewhat different terminology. Mach wrote of facts, instinctive knowledge, and sense perceptions in order to emphasize his belief that the philosophical and psychological basis of science was the immediate experiences of everyday life. Our understanding of planetary motion, for example, ultimately rests on our primal experience with the way heavy objects behave when dropped. To be sure, a lengthy abstract chain of reasoning is required to relate the one experience to the other, but that doesn't mean that the theory is describing the invisible workings of nature. For Mach, the theory was only an economical way of relating experiences, and wasn't itself an element of knowledge. This philosophy successfully rids science of Kantian self-evident truths and metaphysical concepts unrelated to direct experience, but it's unable to cope with the concepts of relativity and quantum mechanics that are fundamentally at odds with ordinary experience.

An ideal Machian explanation of why the pencil balances on the child's finger might be to place the tip of a wire clothes hanger on my finger. The hanger hangs from its tip, just as the pencil does from its, because most of the hanger's weight is below the tip. This explains an unfamiliar phenomenon by showing that it's similar to a familiar one. How effective this approach is in satisfying a four-year-old's need for explanation is unclear, but for a clever adult it might be enough to generate ideas for building other balancing illusions. That is, inventors and mechanics can be very creative working directly with the objects of their immediate experience. Until the rise of modern science in the seventeenth century, this is the way most technological advances were made. And only in the twentieth century have there been technologies, such as nuclear energy and semiconductor devices, that absolutely require theoretical analysis prior to fabrication.

In his last and most important work—*Dialogues Concerning Two New Sciences*—Galileo Galilei (1564-1642) established the foundation for a theoretical understanding of phenomena (Galilei, 1638/1914). Through mathematical comparisons and geometrical deductions, he could explain seemingly contradictory phenomenon. Why, for example, does a piece of chalk take less than a second to fall to the floor, while chalk dust floats in the air, settling to the floor only after many minutes? Imagine, he would say, a piece of chalk subdivided into a million dust particles. Each particle has one millionth (0.000001) the weight of the piece of chalk, but 100 millionths (0.0001) the surface area of the piece of chalk. Thus the air resistance on each dust particle (which is proportional to its area) is 100 times greater, in comparison to its weight, than is the air resistance on a piece of chalk. This explains why the dust falls so much slower than the chalk in terms of very general properties of area and volume.

Newton went beyond Galileo by introducing the invisible force of gravity into his theorizing, justifying it after the fact by the success it gave him in unifying all of mechanics. In those days, a scientific explanation had to be a mechanical one, in which cause and effect were linked, as in a clock, by one part pushing on another. Descartes had tried to explain gravity itself in terms of vortices in the ether, with the result that he could explain nothing. Newton's genius was to postulate as fundamental an attractive force between masses that reached billions of kilometers across empty space. As unmechanical as it sounds, gravity was accepted because it permits calculations of the motion of the planets that are in exact agreement with observation. In fact, gravity soon became part of the "mechanical" picture of the world, although it's as bizarre a notion as any in physics.

Mach accepted universal gravity, because it was somehow connected to primal experience. But he couldn't accept the concept of atoms, because it couldn't be so related. "Atoms cannot be perceived by the senses . . . [and they] are invested with properties that absolutely contradict the attributes hitherto observed in bodies" (Mach, 1942). Such a contradiction violated the principle of continuity in nature, a principle that Mach, the great antimetaphysician, had elevated to metaphysical status. Newton had shown that the force of gravity on earth is continuous with the force on the moon, and ever since science has demanded continuity of its principles. As Mach put it: "That which is a principle of nature in any one time and in any one place, constantly and everywhere recurs, though it may not be with the same prominence" (1942). It appeared to Mach that science had to give up atoms or continuity.

In analyzing these developments, the logical positivists concluded that science could allow explanations based on abstract concepts and general principles, as long as these concepts and principles connect logically to testable consequences. Even the bizarre and unnatural concepts of relativity and quantum mechanics are acceptable since their predictions are ultimately verifiable with the human senses. Philosophically, this is no different than using the balanced pencil to verify the principles that every rigid object has a balance point, called the center of gravity, and that an object will be stably balanced if its center of gravity is below its point of support. Psychologically, the discontinuity of quantum mechanics with primal experience is a special problem only for the learned; students have nearly as much difficulty understanding the "natural" pencil as they do the "unnatural" atom. And at a theoretical level, continuity is preserved even for quantum mechanics, which blends continuously into Newtonian physics as the mass of the object being observed increases (Chapter 3).

In discussing these matters, it's customary to use terms like "experience," "observation," and "observer." Unfortunately, this usage has led some to the mistaken conclusion that science is inherently subjective. If less anthropomorphic words, such as "detectors" and "instrument readings," were used in place of "observer" and "observation," some confusion might be avoided at the expense of a very stilted style. Science is about an objective observer-independent reality, or at least, this is the metaphysical premise of most scientists.

As a child walking to school, I sometimes had the fantasy that all the stores and buildings I passed were merely theatrical sets erected for my benefit. In this fantasy, all my world was indeed a stage and all the people in it were extras hired to populate my life. How could I prove otherwise? Opening a door and finding a store filled with merchandise instead of the back lot of a movie studio couldn't disprove my fantasy, since it's possible that the set designer knew where I would go and had built the set accordingly. Perhaps there is no objective reality after all, and all my experiences are the result of a conspiracy to fool me into thinking that there is one. Such childish thoughts would hardly be worth discussing, were they not the subject of serious philosophical dispute.

How can we know anything, the philosophers declare, if we can't logically distinguish between an objective observer-independent reality and a subjective conspiratorial world? The answer, of course, is that there is no need to distinguish between them. As long as it's assumed that the conspirators perfectly mimic an observer-independent reality, the two premises are logically identical; they are saying the same thing in different words. Upon analysis, the logical positivists found that most traditional philosophical arguments were as empty of substance as this one is (Reichenbach, 1951).

Science is about the interpretation of experience, and since all scientists are human beings, the scientific enterprise is *ipso facto* a human one. Human beings are necessary to interpret the readings of instruments to themselves and to other human beings. It's thus necessary to distinguish between what happens inside a detector (which could be a human head) and the interpretation of these happenings or experiences. The experience doesn't depend on active intellect, whereas interpretation certainly does. In other words, experience is the result of an observer-independent reality, whereas theory is a human construct.

Repeatability

But whose experiences? We all lead very different lives and have very different perceptions of things. How can anything based on something as fragile and variable as human experience be of any value?

Suppose you have constructed the pencil and wire device shown in Fig. 2-1 and balanced the pencil on your finger. Would your experience be the same as mine? Are you doing it at the same time of day as I did? Are you with people or alone? Is there music playing? What kind of music? Clearly your experience, taken as a whole, is different from mine, and from that of anyone else who does the demonstration. Thus, when we use the word "experience" in philosophy we must mean something very different from its everyday meaning. We mean it more in the sense of "experiment," a deliberately isolated and controlled fragment of experience.

Furthermore, this fragment must be repeatable. If you construct the pencil as described, it must balance as promised. This is a critical point, and not as easy as it sounds. Students at Northeastern University do hundred of physics experiments each week, and, I regret to report, they all don't "work." And your pencil may fail to balance.

I have spent a lot of my time analyzing such "failures," so let's look at yours. Perhaps you didn't wrap the wire tightly enough around the pencil, and it is slipping around. Perhaps your wire is thinner than mine, so it doesn't hold its shape. Perhaps your weight isn't heavy enough or your wire isn't long enough. Even this simple device has enough complexity that repeatability isn't assured.

This leads us to a dilemma. If only some pencils balance, then the experience isn't truly repeatable. And if I describe the device in much greater detail in order to increase the percentage of successes—use sixty centimeters (twenty-four inches) of sixteen gauge steel wire with a No. 6 one-hole rubber stopper as a weight—I conceal the fact that the device will actually balance under a very wide range of conditions.

What, then, do we mean by repeatability? Since no two experiences are ever exactly the same in all their details, no two experiences can ever have exactly the same consequences. To achieve repeatability, we must strip away most of the details of our experiences, leaving only a few important factors that different experiences have in common and that determine some important common features of their consequences. This is far from easy. It took Galileo and Christian Huygens nearly a century to discover the factors that determine the period of a pendulum, and this was possible because very few factors are involved and some of them are much more important than others (Matthews, 1994). Thus the effect of the dominant factors could be understood before having to consider the effect of secondary factors.

Oscillating Rod

If a relevant factor is overlooked, experiences don't repeat, as I discovered recently while developing a simple physics experiment in which students carefully measure the period of a steel rod swinging from a pin (Fig. 2-2). The pin is rigidly clamped to a stand and passes through one of three small holes drilled through the rod. Given a slight push, the rod oscillates back and forth for several minutes. By measuring the time taken to execute fifty successive oscillations and dividing this time by fifty, the time for one oscillation, or *period,* can be quickly determined to within two thousandths (0.002) of a second with a stopwatch.

The very simplicity of the system ensured, so I thought, repeatability and agreement with the predictions of Newtonian mechanics. This was indeed the case when the rod was suspended from either of the two holes closest to the end, but not when the rod was suspended from the hole closest to the center. For this third hole, my measurements varied mysteriously from day to day, and none agreed with theory. Although these variations and disagreements were small (about two percent), they were much bigger than the errors in the measurements.

I spent several frustrating months trying to understand this puzzle. In spite of my extensive background in this sort of physics, I couldn't think of anything that would account for a two percent deviation. Then, a long weekend of otherwise erroneous theorizing led me to consider the problem of a ring swinging on a cylinder (Fig. 2-3). On Monday, I went to our machine shop where I scrounged around in the scrap bin until I found an aluminum ring, about six inches in diameter. Back in the laboratory, I immediately started swinging the ring on cylinders of various diameters and, to my complete surprise, found that the period of its swing was very different on cylinders of different diameters.[1]

I had, with the rod, used paper clips of different diameters for the supporting pin, not realizing that such a small change could give measurable differences in the period. The large ring, swinging on large cylinders, enlarged the effect, making its observation inescapable. The lesson is the same as that of the balanced pencil and the cut table-tennis balls of Chapter 1: repeatability requires knowing all the factors that can affect the results. How do we know when we know all the factors? Had I always used the same size paper clips, I would have

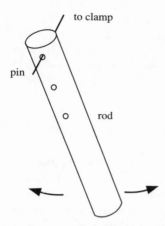

Figure 2-2 A rod oscillating on a pin passing through one of three holes drilled through it.

achieved repeatability without knowing that the size made a difference. However, in describing this experiment to others I wouldn't have mentioned the size of the paper clip (not thinking it relevant). Then, someone else, repeating the experiment with a different size pin, would have discovered the effect. Thus we know we have—if not all the factors, at least the most important ones— when different scientists in different laboratories are able to repeat the experiment.

Explaining the rod in terms of the ring may be said to be a Machian explanation in that it avoids any extraneous concepts. But it isn't very satisfactory. The situation clearly calls for a deeper analysis. Indeed, the results are unpublishable without such analysis. What is required is a mathematical theory that predicts exactly what the period of the ring will be when swinging on cylin-

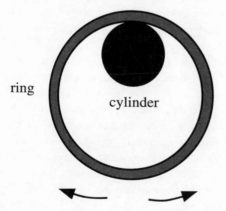

Figure 2-3 A ring swinging back and forth on a stationary cylinder. The time for one swing depends on the radius of the cylinder, as well as the radius of the ring.

ders of different diameters and that correctly predicts what the period of the rod will be when swinging from pins of different diameters. That is, the different sense perceptions must be quantitatively connected by rigorous mathematical analysis. Only when this is achieved, can one say with some satisfaction that one has explained the phenomena (Cromer, 1995).

Hempel Analogy

Modern philosophy interprets scientific knowledge as the connection of theory and experience, a peculiar marriage of metaphysics and empiricism. The philosopher Carl Hempel likened scientific theory to a net suspended over the world of experience (observations and experiments) (Fig. 2-4). The knots of the net are the terms and the threads are the relations—definitions and laws—among the terms. The net is connected to experiences by rules of interpretation, which he likened to strings running from points on the net to experiences in the world below. An experience in one place is related to an experience in another by going up to the net along the string at the first experience, traveling along the net to the string attached to the second experience, and going down this string (Carnap, 1966). In this view, the theoretical framework is not just an economical device for relating different experiences, but it's an indispensable part of a connected whole.

A scientific theory must specify the relevant factors of an experience and how changes in these factors change the outcome of the experience. In terms of the Hempel diagram, the theory logically connects many different specific experiences, so that they are equivalent from the point of view of repeatability. The rigid rod is an excellent example of this.

From Newtonian theory one can derive a formula for the period of the rod when swinging from pins through holes at different distances from the end of the rod. The theory is exact, but the formula is an approximation, because it doesn't includes all the relevant factors. Such an approximation may be called a "model," because, as a model ship, it represents some relevant factors. A purist could argue that Newtonian mechanics is also a model, since it isn't truly exact: it excludes the effect of special relativity. But this effect is a correction to a rod's period that's billions of times smaller than we could ever hope to measure, so the inexactness of Newtonian mechanics isn't relevant here.[2]

In the simplest case, the formula, or model, for the period of the oscillating rod includes the length of the rod and the distance of the suspension hole from the end, but not the mass or diameter of the rod. This means that the model applies to rods of different lengths, hung from holes at different distances from their ends, and made of different materials. To repeat my experiments, you don't need my particular rod, or even one of the same dimensions or material. Your results, like mine, would agree within 0.5 percent with the predictions of the model for holes close to the end, but would disagree by several percent for holes close to the center.

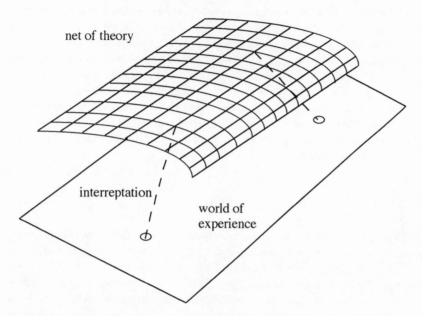

Figure 2-4 Hempel's net of theory, suspended over the world of experience.

A better model would include, as additional factors, the radius of the hole and the diameter of the pin through it. This would bring all experiments into agreement with the model at the 0.5 percent level. A still better model would include still more factors, such as the diameter of the rod, the angle through which the rod swings, and the viscosity of air. With each refinement, ever more diverse experiences can be considered repeats of one another for the purpose of testing the model, and, indirectly, the theory.

There are two points that relate this philosophy directly to education. First, one can do very little science if one stays in the world of observation alone. Students must do some traveling between observations and theory along the connecting strings in order to experience real science. Second, the threads of the theoretical net aren't directly observable. Thus, there's no possibility of students discovering them by experimentation, as constructivist educators seem to believe. Students must be taught the theory, using demonstrations, experiments, guided inquiry, and traditional problem-solving methods. Although guided inquiry is often called constructivist, it assumes, as does all traditional education, that there is something that teachers know and that their students have to learn.

The best example of a nonobservable thread is the mother of all threads, the law of inertia, or Newton's first law of motion: "Every body continues in its state of rest, or uniform motion in a right [straight] line, unless it is compelled to change that state by forces impressed upon it" (Newton 1686/1934). Earlier in the *Principia,* Newton defined an impressed force as "an action exerted upon a body, in order to change its state, either of rest, or of uniform motion in a right line," so the law of inertia is inherently circular. Furthermore, the law is untestable, since there is no way to observe a body without forces "impressed upon it" moving forever in a straight line with constant speed. This didn't bother Newton, who tells us that "in philosophical disquisitions, we ought to abstract from our senses, and consider things themselves, distinct from what are only sensible measures of them." That is, the fundamental principles of the theory must be obtained by a process of abstraction from observations; they aren't themselves observable.

SEED

To help students with this abstraction process, teachers must provide their students with relevant demonstrations and experiments. SEED begins the discussion of Newton's first law by pushing a book across a table. When the pushing stops, the book stops moving. "See," SEED says, "it's obvious that a force is needed to keep an object in motion; Aristotle was right. Or was he?" The book is then placed on a layer of drinking straws and again pushed; now it continues to glide across the straws. "What's going on?" SEED asks. After some discussion, it's understood that decreasing the opposing force (friction) allowed the book to continue moving for a short time after the push stops. "What would happen if the friction were reduced further?" More discussion. A demonstration follows in which an air puck, floating on a cushion of air supplied by an inflated balloon, is given a little push. The puck then glides across the table with no apparent decrease in its speed.

In many respects, the SEED approach is close to the spirit of Galileo in his *Dialogues on Two New Sciences.* SEED's use of simple materials, suitable for middle-school use, is in keeping with Galileo's belief that it's the mundane phenomena of everyday life that best reveal the laws of nature.

> We come now to the other questions, relating to pendulums, a subject which may appear to many exceedingly arid, especially to those philosophers who are continually occupied with the more profound questions of nature. Nevertheless, the problem is one which I do not scorn. I am encouraged by the example of Aristotle whom I admire because he did not fail to discuss every subject which he thought in any degree worthy of consideration. (Galilei, 1638/1914)

SEED used simple materials initially because the teachers don't have the money for real equipment. However, we've found that with familiar materials, the phenomena stand out more as aspects of nature and not as peculiarities of particular pieces of apparatus. For this reason, a number of simple SEED-like experiments have been introduced into the college-level physics laboratory.

But the core of any science instruction must be to develop some understanding of the theoretical structure of the subject and how it relates to observation. Galileo showed how this could be done, introducing many of the examples which are still part of the canon of physical science: scaling of area and volume, the period of a pendulum, the frequency of a vibrating string. In all cases, measurement is the connecting link between observation and theory. Starting with the measurement of length, students can, through repeated exercises, develop an intimate understanding of quantity, estimation, interpolation, and uncertainty and apply these concepts and skills to theoretical problems. Unfortunately, few teachers have any experience with this level of scientific inquiry.

Golden Ratio

For example, I was told when I was in school, and students are still told today, that the ratio of the length to width of the most aesthetically pleasing rectangle is a special number, called the "golden ratio," or $\phi = 1.618. \ldots$ It is also said that ϕ is the ratio of certain bones in the human body, or is the ratio of the width to height of a butterfly's wingspan. That none of these statements is true isn't the problem. They are all easily testable hypotheses which can interest students and engage them in the important exercise of measuring lengths and calculating ratios. The problem is that teachers seldom approach the statements this way.

In a special ten-week enrichment program for gifted fifth graders, a teacher had her students do poster exhibits on the golden ratio. Some had pictures of the Parthenon, with the usual statements about its proportions having the golden ratio, and others had pictures of butterflies with their wing spans inscribed in a rectangle whose sides were reputed to be in the golden ratio. The teacher had worked hard to develop an interesting activity with great potential for applying mathematics to science. But none of the potential was realized. In ten weeks, the students never made any measurements of butterflies, either real or pictured, to check whether the length divided by the width of the butterflies' wing spans was in fact equal to ϕ. When asked about this, these bright fifth graders responded that they took the teacher's word for it. All that time, all that opportunity, wasted. Worse than wasted; the students were rewarded for exhibiting mindless recitation of dubious facts that they could, and should, have checked for themselves.

The very process of making measurement in order to check a theoretical claim is very empowering for students and teachers alike. It was a revolutionary concept when developed by Galileo and Newton, and is still unfamiliar to most elementary and middle-school science teachers. It doesn't come naturally; it's an attitude that develops only after years of studying and teaching experimental science.

The statement "Phi (φ) occurs frequently in nature" is a good example of a possible law of nature that students can test and philosophers can discuss. I'm not worried about the vagueness of "occurs frequently," since the whole statement is to be interpreted in the same sense as the statement "Pi (π) occurs frequently in nature." Pi (π = 3.14159 . . .), the ratio of the circumference of a circle to its diameter, occurs in many formulas, such as the period of Galileo's pendulum, the period of my oscillating rod, and the probability that a paper plate, scattered randomly on a tiled floor, will land on a corner (Cromer and Zahopoulos, 1993). The golden ratio also has has many interesting mathematical properties,[3] but, unlike π, it isn't important in physics (Gardner, 1994). There is thus no physical reason for art or nature to favor the golden ratio, which makes it difficult to establish the claim that they do.

For example, even an exhaustive study that showed that the ratio of the length to width of the wing spans of all butterflies was between 1.60 and 1.64 wouldn't, by itself, prove that this ratio is φ. The actual ratio could be 8/5 (= 1.6), or 81/50 (= 1.62), or completely random within a narrow range. However, if there was some theory that predicted on some rational basis that the ratio should be φ, then a range of values between 1.60 and 1.64 would be supportive, though not conclusive.

But judging from a few measurements I made on pictures of butterflies, the ratio actually varies between 1.4 and 2.4. Thus fifth graders could have easily disproved the hypothesis, since it has neither theoretical or empirical justification. It's also easy to show that the golden rectangle isn't judged to be the most aesthetic rectangle (Markowsky, 1992). The golden ratio, although a fascinating number in mathematics, is a red herring in nature, and this should be the conclusion of any honest study of it in a science class.

Theory and Observation

Science requires a theoretical structure prior to experimentation and measurement, since without a theory it isn't possible to extract from the multitude of factors present in every situation those that are relevant. Some of this structure is so general that it's taken for granted. For example, we assume that the period of a pendulum doesn't depend on who makes the measurement, or when or where the measurement is made. On this assumption, data taken by different students on different days in different classrooms can be compared and combined.

The theory-laden nature of science, critics say, robs science of its claim to objectivity. Science sees only what it looks for, and it looks for only what its theory tells it to look for. Science is thus a conventional enterprise, no different in structure from theology or literary criticism. Its practitioners share common conventions and beliefs, and only undertake activities that reinforce their prejudices, thus giving them a false sense of the objectivity of their work.

Scientists like to tell the story of William Crookes, the nineteenth-century inventor of the cathode-ray (electron-beam) tube, who, finding that some unexposed photographic film in his laboratory was fogged, sent it back to the manufacturer. The point is that he failed to see the significance of the fogging, because it wasn't part of his theoretical understanding. Of course, Wilhelm Roentgen in 1895 did investigate the faint glow that he had noticed coming from a fluorescent screen near his covered cathode-ray tube, and discovered X rays (Farmelo, 1995). Within a year, X rays were being used to diagnose fractures and locate bullets in wounds, and they've been an indispensable part of science and medicine ever since. That another scientist failed to investigate a mysterious fogging of his photographic film doesn't matter much. It doesn't matter how many scientists fail to follow up an important clue, as long as one does.

Kuhn makes the distinction between normal science and revolutionary science. But Roentgen's great discovery, which has had revolutionary consequences for medicine and physics, was part and parcel of normal science. Roentgen was performing normal electron-beam experiments and found something unusual. The unusual is part of normal science—it's what every scientist dreams of running across. The great fear is that when fate throws you the winning pass, you'll fumble it. But it's also normal for some scientists to make the catch, as Roentgen did.

It should also be added that X rays, as epochal as they were, fit into the general theory of electromagnetic radiation that had been developed by James Clerk Maxwell. By 1895, radio waves, infrared radiation, light, and ultraviolet radiation were known to be electromagnetic radiations differing only in frequency. X rays fit onto the high-frequency end of this continuum, beyond the ultraviolet.

Our confidence in the objective reality of X rays is based jointly on the experimental evidence and the theoretical framework of Maxwell's equations of electromagnetism. This is in stark contrast to cold fusion, which has neither. Of course, it's possible, in principle, to accept the observational facts of X rays, while denying their ontological reality. But this is just philosophical word-twisting which would as well deny the reality of the chair you're sitting on. More reasonable, philosophically, is the position that Michael Matthews has called "modest realism": acceptance of an independent external reality that scientific theories describe with ever-improving accuracy (Matthews, 1994).[4]

Soon after X rays were discovered, crystals were placed between the cathode-ray tube and photographic film. The orderly array of the atoms in the crystals produced an ordered array of dots on the film. From an analysis of the pattern of dots, the exact arrangement of the atoms in the crystal, including their spacing, could be deduced. This technique was applied to ever more complex crystals until, in the early 1950s, it was applied to crystallized DNA extracted from the cells of living organisms. From the X ray work of Rosalind Franklin, James Watson and Francis Crick were able to deduce the structure of this molecule, revealing the molecular basis for reproduction and heredity (Watson, 1968). Thus, the nineteenth-century observations of Roentgen on cathode-rays and of Gregor Mendel on sweet peas are connected to each other through a vast overarching theoretical network.

The objectivity of science isn't established by any one observation or theory, but by the totality of its theories and observations. Although we'd like our students to acquire a unified understanding of science, it's doubtful that it can be taught this way. Certainly the new science textbooks discussed in Chapter 1 fail to do so. Their authors, who have no deep understanding of science themselves, have bitten off subjects that are too large for them, or their readers, to chew. The integrated science curricula urged by reformers will fail because there are very few teachers with a sufficiently broad science background to teach them in any meaningful way.

It's more practical, I believe, to concentrate, *a la* Galileo, on certain self-contained subjects—such as statics, simple machines, electric circuits, optics—that have relatively simple theoretical structures which students can connect to observations. Laboratory work is essential here, for students to experience the feel of objectivity. The size and complexity of the subjects can grow with the age and sophistication of the student, eventually connecting (say) optics with perception and statics with biomechanics. But any notion that students must build their knowledge in an orderly fashion, from the simpler to the more complex, is totally missing from the recent textbooks.

Indeed, the 1994 draft of *National Science Education Standards* (National Research Council, 1994) said that science teachers can skip years of study, yet gain depth of understanding, if they are provided with "opportunities for inquiry." Suggested opportunities were "full inquiry into the genetic variation of 'fast plants,' intensive laboratory work in visualizing sound waves, or systematic observation of a gull community on Lake Erie." There was no appreciation that these are full-time research projects and that a student would need specialized knowledge in genetics, physics, or ethology, as the case may be, before any meaningful "inquiry" would be possible.

What we have here is a dispute between a holistic or "big-picture" approach to science—and to education in general—and an analytic or "first-things-first" approach. My position—that one has to understand the little problems before one can tackle the big ones—is very unpopular in educational circles these days. But as the public grows weary of the continuing failure of holistic education to produce a literate population, alternative approaches will have to be considered.

Thickness of Paper

The first investigation in the introductory physics laboratory at Northeastern University illustrates how scientific objectivity can be taught through a careful analysis of a little problem. The problem is just to measure the thickness of a page in a book using a plastic ruler graduated in millimeters (1 mm = 0.04 in.). Specifically, they are asked to measure as carefully as they can the thickness of one, ten, twenty, forty, seventy, 100, and 150 sheets in a book, and to estimate the uncertainty in each measurement (Cromer, 1994). For example, the thickness of seventy sheets (140 pages) might be found to be 6.0 mm. The uncertainty in reading a ruler is half the distance between its divisions, or 0.5 mm in this case.

Assuming that the thickness of seventy sheets is just seventy times the thickness of one sheet, the thickness of one sheet is (6.0 ÷ 70) = 0.086 mm with an uncertainty of (0.5 ÷ 70) = 0.007 mm. This result is written 0.086±0.007 mm, meaning that the thickness is most likely between 0.079 and 0.093 mm.

Students aren't too impressed with this. That is, they don't really believe they are measuring the thickness of a piece of paper. It doesn't have the feel of objectivity. To get this sense of a real measurement it's necessary to compare the results obtained by different groups of students. This is a critical part of the laboratory experience that we're only starting to emphasize.

The student laboratories at Northeastern are taught by physics graduate students, who first do all the experiments under faculty supervision. To show new graduate students the value of comparing results, I have them put their own data on the board. Table 2-1 shows the results of the thickness investigation obtained by four groups of first-year graduate students from measurements of the thickness of seventy and 150 sheets. The number before each plus-minus sign (±) is the thickness of a sheet obtained by dividing the measured thickness of a stack by the number of sheets in it. The number after each plus-minus sign is the *estimated* uncertainty obtained by dividing 0.5 mm by seventy and 150, as appropriate. Although a 150-sheet stack is proportionally thicker than a seventy-sheet stack, the uncertainty in reading a ruler doesn't change with the thickness of the stack, and so the estimated uncertainty per sheet decreases with the number of sheets.

The *real* uncertainty is the spread of the values obtained by the different groups. Groups A, B, and C found values for the thickness which agree with one another within their estimated uncertainty. Furthermore, the spread in their values is smaller with 150 sheets than with seventy sheets, as expected from the smaller estimated uncertainty. The convergence of the measurements

Table 2-1

Thickness of a Sheet of Paper, in Millimeters, Measured by Four Groups of Physics Graduate Students.

Group	Thickness of a sheet of paper calculated from measurements of the thickness of 70 sheets	Thickness of a sheet of paper calculated from measurements of the thickness of 150 sheets
A	0.086±0.007	0.083±0.003
B	0.082±0.007	0.086±0.003
C	0.089±0.007	0.082±0.003
D	0.102±0.007	0.097±0.003

Note: The number before each plus-minus sign (±) is the the thickness of a sheet obtained by dividing the measured thickness of the stack by the number of sheets in it, and the number after the plus-minus sign is the estimated uncertainty obtained by dividing 0.5 mm by the number of sheets in the stack.

of groups A, B, and C to a value around 0.084±0.003 mm begins to have the feel of objectivity.

The marked discrepancy of group D's values suggest that the group either used a procedure different from that used by groups A, B, and C, or that it was measuring the pages in a different book. Had the former been the case, a discussion would have ensued to decide which was the better procedure. As it turned out, group D had measured a different book with a different grade of paper.

But even this didn't convince all the graduate students that they had really measured the thickness of a piece of paper. One asked how we know that all the sheets in a book have the same thickness. Another asked whether there might be some space between the sheets, and consequently, whether the result would change if the sheets were tightly pressed together. These skeptics wanted to measure the thickness of a single sheet of paper with a micrometer, to check that the procedure of measuring 150 sheets with a plastic ruler really does measure the thickness of a single sheet.

This discussion shows that to achieve consensus about the meaning of a procedure, it's necessary to obtain the same result by alternative means. No matter how consistent the measurements are using multiple layers of paper, some will claim that it's just an arbitrary convention to call the resulting number, say 0.084±0.003 mm, the thickness of a sheet of paper, just as some claim that it's just an arbitrary convention to call the score on certain tests a measure of intelligence. Objectivity requires not only repeatability, but validation. With a micrometer measurement of 0.088±0.004 mm and a microscope measurement of 0.085±0.002 mm, almost everyone would be convinced that "the thickness of a sheet of paper" is a meaningful concept and that for the particular paper in question its value is 0.085 mm, give or take a few thousandths of a millimeter.

Conclusion

Repeatability is the essence of scientific knowledge, yet it is possible only if there is prior theoretical knowledge concerning the relevant factors to be repeated. Even the simple process of measuring with a ruler has a theoretical basis; it's not intuitive. But since theory must be based on repeatable experiences, the whole process of science appears to be circular. Actually, it's more spiral, the cyclical exchange between theory and experience lifting and refining the scope and precision of the concepts and methods. Such a process is hard to get going. There are many theories that just don't make it. What survives has met stringent tests of repeatability, consistency, and validity, and can, without blushing, be called objective scientific knowledge.

Postmodern constructivists see science as a purely empirical endeavor. For them, science is like the low stone boundary walls that New England farmers built from stones extracted from the rocky soil and which now run through woods that long ago reclaimed the harsh farmlands. No mortar was used in constructing these walls, but some care was taken to fit smaller stones in between larger ones to create a weakly interconnected structure which might be hundreds of feet long and seldom more than two feet high. The stones, like the facts of pure empiricism, support only a few of their neighbors, so that a bulldozer, pushing through a six-foot-wide section of the wall, leaves the rest of the wall unaffected.

Science, in contrast, can be likened to a stone arch, soaring into the sky. Each carefully shaped stone is essential for the support of all the others. Remove one, and the entire structure collapses. But leave them alone, and the arch is strong enough to support the roads and aqueducts of an empire. And like science, the arch, once built, no longer requires the rickety scaffolding used in its construction. Some knowledge of the construction process is helpful in understanding how such a self-supporting structure came into being, yet whatever its history, the structure itself is eternal.

Critics of objectivity might complain that by basing my arguments on simple mechanical examples, such as the balancing pencil, the oscillating rod, and the thickness of a piece of paper, I have ignored the deeper philosophical issue raised by modern science. Doesn't quantum mechanics deny the very causal and deterministic assumptions that underlie all mechanical notions of reality? Doesn't quantum mechanics tell us that the act of observation unavoidably affects the outcome of the observation, eliminating the very foundation of an observer-independent reality?

These issues are addressed in Chapter 3. The discussion is necessarily technical because physics at the atomic level, as described by quantum mechanics, doesn't correspond to one's primal notions of how things should be. Its uncertainty principle violates mechanical intuition, its intrinsic statistical nature violates determinism, and its paradoxical consequences violate common sense. Yet, quantum mechanics is undoubtedly the greatest triumph of twentieth-century physics, and seventy years after its development, it remains unchanged, if not unchallenged. The essential conclusion of Chapter 3, for those who choose to skip it, is that, contrary to popular opinion, quantum mechanics insures the certainty of scientific knowledge.

3

Certainty and Uncertainty

The Gilbert chemistry set I got for Christmas when I was seven years old was put away until my next birthday because the box said it was for children ages eight and older. The wait was painful, but my reading improved so much during this period that when I did open the box, I could read the instruction manual well enough to do simple experiments. I made clear solutions turn pink, popped corks out of bottles with vinegar and baking soda, and created precipitates by mixing a cobalt chloride solution with various salt solutions. The pleasure I got has remained with me all my life, and I incorporated some of these same experiments into the SEED program so that I could get paid to do what I did for fun when I was eight years old.

My sister was taking college chemistry at the time, and she taught me about atoms and molecules, using cut-out paper atoms that came with the chemistry set. The hydrogen atom was in the shape of a small circle with an arm that was pointed at its end; the oxygen atom was a larger circle with two arms that were notched at their ends (Fig. 3-1). Thus, two hydrogen atoms could be fitted into one oxygen atom to form a molecule of water. From these ingenious models, I quickly learned about the valences of the atoms—it was, after all, child's play.

There were, of course, one or two things I didn't quite understand. For instance, I couldn't picture how the atoms really hooked together. I tried picturing a hook-and-eye mechanism in which the hook of a positive valence fastened into the eye of a negative valence. This was fine, but what were the hooks and eyes made of? I had been told that atoms were the smallest objects in the world and that everything else was combinations of them. Were the hooks and eyes of the atoms made of still smaller entities—atoms within atoms, so to speak? And if so, how did these smaller entities fasten together to make the hooks and eyes of the atoms? Where they made of still smaller particles, and so *ad infinitum*?

These simple questions don't have simple answers. In fact, real understanding of the chemical bond—the mechanism that holds atoms together—had emerged only in the decade or so preceding these musings. And many—

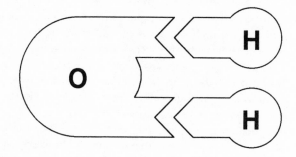

Figure 3-1 Cut-out atoms from a Gilbert chemistry set, ca. 1943.

including Einstein—expressed surprise at the strange indeterministic physics that lay at the heart of the atom, though it was clear to a child that atoms couldn't be made of stuff the way stuff is made of atoms.

Birth of the Quantum

By 1890, almost all known physical phenomena had been successfully explained in terms of Newton's mechanics and Maxwell's electromagnetism. All of optics was shown to be a consequence of the interaction of electromagnetic waves at the boundary of different transparent materials, and the second law of thermodynamics was shown to be a consequence of the statistical behavior of innumerable atoms moving in accordance with Newtonian mechanics. This grand system of theories we now call "classical physics," to distinguish it from the "modern physics" of relativity and quantum mechanics that developed between 1900 and 1925.

The astounding theoretical successes and philosophical conundrums of modern physics have been the main focus of historians and philosophers of science in this century, often obscuring the fact that classical physics was almost totally successful within its own domain, and it remains so to this day. By and large, modern physics replaced classical physics only when describing the internal structure of atoms.

However, there were several important phenomena, within the domain of its theories, that classical physics failed to explain. One of these was the spectral distribution of the light emitted by luminous objects. You can study this phenomenon qualitatively if you have a dimmer switch controlling an incandescent light bulb. With the dimmer turned to low, no visible light is given off, but as the dimmer is turned up slightly, the filament in the bulb will give off a faint red glow. Turning up the dimmer a little more, the light given off by the filament becomes brighter and more yellow. Continuing this way, you will see the color of the light change as it gets brighter.

If light from the bulb were passed through a prism, it would be separated into a rainbow of colors. The relative brightness of these colors would change with the brightness of the bulb. The dim red light from the bulb would have mostly red in its spectrum, with little yellow and even less blue. As the bulb glowed more brightly, its spectrum would have increasing proportions of yellow and blue. Instruments sensitive to radiation with frequencies lower than red light (infrared radiation) and higher than blue light (ultraviolet radiation) would detect small amounts of these radiations as well.

Studies of this phenomenon in the nineteenth century showed that all solid bodies are constantly emitting a distribution of radiation similar to what we see from a light bulb. For cooler objects, most of the radiation is low-frequency infrared and microwave radiation, which is invisible to the unaided eye. However, regardless of the temperature, the distribution has the characteristic shape shown by the "experimental result" curve in Fig. 3-2. This curve peaks at a frequency that depends on temperature, and falls off rapidly on either side of the peak. At sufficiently high temperature, there is enough radiation in the visible frequency range to be seen.

The shape of the distribution curve depends to some extent on the reflective properties of the emitting body. However, all perfectly nonreflecting bodies—so-called "black bodies"—have the same spectra at the same temperature. The most perfect black body on earth is a small hole in a large container, since very little light entering the hole is reflected back out. In the nineteenth century, spectroscopists measured the spectrum of the black-body radiation coming from such holes for different temperatures of the container, obtaining the "experimental result" shown in Fig. 3-2.

In 1965, A. A. Penzias and R. W. Wilson discovered that outer space itself was a black body with a temperature of −270° C, only 2.7° C above absolute zero. At this low temperature, the radiation is in the microwave region of the electromagnetic spectrum. The spectrum of this radiation, which is believed to be left over from the early expansion of the universe, has been measured with a precision of one part in a million.

Thus, one of the best established facts of nature is that solid objects emit a particular lopsided bell-shaped distribution of radiation that has its peak at a frequency that increases with the temperature of the object. There is nothing strange about this. In fact, the speed of the molecules in a gas have the same type of distribution, with a peak at an energy that increases with the temperature of the gas. One of the brilliant successes of classical physics was its ability to predict this distribution exactly on the basis of sophisticated statistical analysis. But when this same theoretical analysis is applied to the distribution of frequencies in black-body radiation, a peak isn't deduced. Instead, the amount of radiation just continues to increase with frequency, as indicated by the "classical prediction" curve in Fig. 3-2. Since there is no highest frequency, classical physics predicts that a body, at any temperature, would emit an infinite amount of radiation—an absurd result.

The essential conclusion is that radiation and molecules differ less in their behavior than classical theory predicts that they should. Although a gas appears

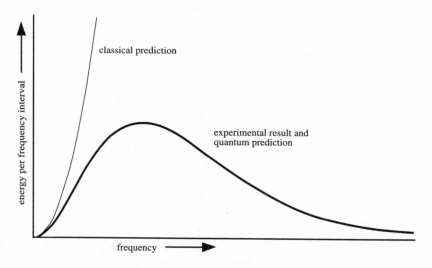

Figure 3-2 The distribution of black-body radiation. The prediction of quantum theory coincides with the experimental data, whereas the prediction of classical physics agrees with the data only at the low-frequency end of the distributions.

to us to be a continuous substance, it's actually composed of an immense number of individual molecules. There are, for example, 10^{22} molecules in a liter of air. Yet as large as this number is, it is finite. In classical physics, radiation is considered to be truly continuous; there isn't a smallest bit of red light.

In 1900, Max Planck showed that the experimental black-body radiation distribution could be derived from basic theory if radiation was treated as though it too had some degree of graininess. He hypothesized that whenever radiation interacted with matter it gained or lost energy in discrete units, or quanta, proportional to its frequency. With only one new number—Planck's constant h—Planck obtained a lopsided bell-shaped distribution that exactly matched the experimental one. This triumph, as revolutionary as it proved to be, was a product of normal science. It takes nothing away from Planck's creative genius to point out that he worked within the highly developed theoretical framework of his time and that he wasn't alone in his quest for an understanding of black-body radiation. The laurel wreath goes to him for winning a race that, sooner or later, someone was bound to win.

In 1905, Albert Einstein took Planck's hypothesis one step further, saying that radiation itself is composed of discrete units of energy, called "photons" and that the energy of a photon of frequency f is hf. This explained the "photoelectric effect," a recently discovered phenomenon in which blue light, no matter how faint, can eject electrons from some metals, whereas red light, no matter how bright, can't. Each photon of blue light has a greater energy than each photon of red light because the frequency of blue light is greater than the frequency of red light. A single blue photon has enough energy to eject a single electron from the metal, whereas a single red photon doesn't. It doesn't matter how many red photons fall on the metal, because each is insufficient to its task and two photons almost never interact with the same electron at the same time.

With Einstein's photon theory, radiation looks very similar to a gas. Photons with energies much greater than average are rare, just as are air molecules with energies much greater than the average. Through the Planck-Einstein relation $E = hf$, we understand the absence of high-frequency radiation in the blackbody distribution as just the absence of high-energy photons.

Radiation Hazard

There is much concern, and much misunderstanding, about the potential harmful effects of electromagnetic radiation of different frequencies. All matter, including living matter, is composed of atoms that are bound together by electrons. A photon can disrupt a bond only if it has enough energy to knock out an electron. Photons of visible light don't have this much energy, but they are readily absorbed by pigment molecules in the skin. Higher-frequency radiation is not as readily absorbed, and matter becomes increasingly transparent to radiation at ultraviolet, X-ray, and gamma-ray frequencies. On the other hand, the energy of each photon is hundreds, thousands, and in the case of gamma radiation, millions of times greater than the energy of a blue photon. So on the rare occasion that a photon of (say) X-ray radiation does strike an electron, the electron will be knocked out of its atom and the atom may disconnect from its molecule. Furthermore, the knocked-out electron has sufficient energy to break up many other molecules. Should one be a particular segment of DNA in a living cell, the result could change the cell from normal to cancerous. For this reason, all radiation with frequencies greater than light are considered biologically hazardous.[1]

Radiation with frequencies below that of light have correspondingly lower energies. For infrared radiation, each photon has one-hundredth the energy of a blue photon, for microwave radiation, each photon has one hundred-thousandth the energy, and for the very low frequency radiation generated by sixty-cycle electrical power lines, each photon has one-trillionth the energy. For this reason, all radiation with frequencies less than light are considered biologically nonhazardous, except at intensities high enough to cause heating.[2]

In spite of these facts, which have been known since 1905, there has been great uproar over the possible dangers of very low levels of very low-frequency radiation. In 1992, a Swedish study (Feychting and Ahlbom, 1993) was widely reported to have demonstrated a connection between proximity to high-voltage power lines and leukemia in children (Gorman, 1992). It was reported to be the best study to date, since it analyzed a registry of close to half a million individuals who had lived within 300 meters of any of Sweden's high-voltage lines between 1960 and 1985. The population was divided by nearness to the power lines and this distance was used as a measure of dosage. However, distance didn't correlate with the levels of radiation actually measured by the investigators in randomly selected houses.

Furthermore, there was no difference in the cancer rates of children living different distances from the power line. To get a difference, the investigators had to look at a subset of childhood cancers. They found that leukemia had a higher rate for children who had lived near the power line, though how long they had lived near the line, or whether the distance properly determined their exposure, wasn't established. There were twenty-seven cases of childhood leukemia in the group farthest from the lines. The group nearest to the lines had one-fifth the population and so it was expected to have five or six cases, whereas it actually had eleven. But of all thirty-eight cases, only eleven had lived at the same address from birth to the time of diagnosis. Of these eleven cases, only two had lived near the line, which is just the statistically expected number. In spite of all the talk about a population of half a million, the study came down to a handful of dubious cases. The laws of physics can't be overthrown by studies like this, but the public can be panicked into demanding costly remedies to a nonexistent danger.[3]

Bohr Atom

The quantum concept brilliantly explains the distribution of the frequencies of radiation that are emitted from a black body as well as the distribution of the speeds of the electrons emitted from a metal that's illuminated by light of different frequencies. But the quantum relation $E = hf$ equates logically inconsistent quantities. The energy E is of a discrete localized entity, a photon, whereas the frequency f is of a continuous nonlocalized entity, a wave. As a matter of indisputable experimental fact, radiation exhibits both quantum and classical properties

The greatest success of the quantum concept in this period came when Niels Bohr used it to explain the structure of the hydrogen atom. The year was 1913, just sixteen years after Joseph J. Thomson discovered the electron. Experiments with so-called cathode rays had been conducted throughout the 1890s; Roentgen's cathode-ray experiments led to the discovery of X rays in 1895 (Chapter 2).

A cathode-ray tube is a partially evacuated glass tube that has two wires passing into it (Fig. 3-3). When these wires are connected to a high-voltage source, the tube glows mysteriously. The electrical circuit is somehow being completed by the flow of something through the near vacuum of the tube. A cathode-ray tube used for physics lecture-demonstrations has a fluorescent screen sealed inside it. A glowing blue-green line appears on the screen as the something streams along side it. This line can be bent up and down with a U-shaped magnet oriented one way or another around the tube, which demonstrates that the something is charged. By measuring the degree of bending for

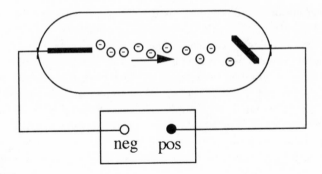

Figure 3-3 Cathode-ray tube, consisting of a partially evacuated glass tube with two wires in it. When these wires are connected to a high-voltage source, electrons flow from the negative wire to the positive wire.

a known magnetic field, Thomson determined that the something was a beam of negatively charged particles with a definite mass. This particle he called the electron.

You can easily repeat the qualitative aspects of Thomson's experiment at home, because every television picture tube is a cathode-ray tube. The electron beam sweeps across the face of the tube, which is coated with fluorescent materials that glow with different colors when struck by electrons. If you take a small magnet from the side of your refrigerator and hold it close to a television screen, you will produce a color change in the picture near the magnet. The magnet exerts a force on the moving electrons, causing them to miss their intended targets. With a stronger magnet, the picture itself will be distorted.[4]

From Thomson's work, it was determined that the electron had a mass much less than that of an atom. It was reasonable to assume that the bulk of an atom's mass, which had to be positively charged to neutralize the negative charge of the electron, entirely filled the atom. But experiments conducted in 1911 showed that this mass was concentrated in an extremely small region of space. Thus a picture emerged of an atom as a collection of very light negatively charged electrons swarming about a tiny, massive positively charged nucleus. The analogy with the solar system is very close, especially as the attractive electrical force between electrons and their nucleus decreases with distance in the same way as does the gravitational force between planets and the sun.

Thus, it would seem that the entire machinery of Newtonian mechanics, developed originally to explain the dynamics of the solar system, could be applied to the dynamics of the atom. Except that it can't be. The fundamental difference between atoms and solar systems is that all atoms of the same element are identical, whereas all solar systems are different.

The earth happens to have a certain orbit about the sun, but as far as Newtonian mechanics is concerned, it could have had any other. In any other planetary system, the planets would certainly orbit at different distances and at different speeds than do the planets in our system. Each planetary system is the result of its own peculiar history. From the positions and velocities of all the masses in a system at any one time, Newtonian mechanics determines their positions and velocities at any other time. But it doesn't say what the positions and velocities are in the first place. These must be found from observation. Newtonian mechanics is deterministic in that future positions and velocities are determined from past positions and velocities, but it isn't prescriptive; it determines what will happen given certain facts, but it doesn't prescribe these facts.

Jumping from the success of Newtonian mechanics in predicting the motion of a few planets and comets around the sun, the French astronomer-mathematician Pierre-Simon Laplace (1749–1827) declared that an intelligence who knew the positions and velocities of all the particles in the universe at one time could predict their positions and velocities at any other time, future or past. This strict form of determinism is generally thought to be quite incompatible with any notion of free will, and leads to the familiar Laplacian defense plea: "Your honor, I'm not responsible for what I did, since it was determined from the positions and velocities of the atoms at creation that I would do it." To which the Laplacian judge responds; "True, but it was also determined from the positions and velocities of the atoms at creation that I would order you to be hanged."

Newtonian mechanics relates the future to the past, giving us historical, rather than absolute, determinism. According to Newtonian mechanics, the orbit of the solitary electron in a hydrogen atom should change as the atom collides with other atoms, and at any instant reflect the atom's unique history. But all hydrogen atoms are absolutely identical under normal conditions, a fact that's inexplicable in Newtonian mechanics.

In order to understand the non-Newtonian behavior of atoms, Niels Bohr developed a hybrid theory that mixed Newtonian mechanics with quantization. He hypothesized that for some as yet unknown reason the electron in a hydrogen atom, although behaving like a Newtonian particle in all other respects, can only move in orbits in which its angular momentum is a whole multiple of $h/2\pi$. From the continuous infinity of possible orbits allowed by Newtonian mechanics, the Bohr quantum condition permits only a discrete infinity; one orbit for each integer 1, 2, 3, ... This integer is called the "quantum number."

The size of each allowed orbit is *determined* by Planck's constant h, the mass and charge of the electron, and the quantum number; all references to initial conditions are eliminated in this theory. Furthermore, the smallest quantum number, *numero uno,* corresponds to the orbit, or "state," with the lowest energy. An electron in the lowest, or "ground," state of hydrogen can jump into the next state, with quantum number two, only if it absorbs a quantity of energy corresponding to a photon of ultraviolet radiation. This is considerably more energy than a hydrogen atom receives in any single collision with other atoms at room temperature, so under normal conditions, all atoms remain in their ground state.

However, if a small amount of hydrogen gas is put into a cathode-ray tube, where they are bombarded by energetic electrons flowing between the wires, some hydrogen-atom electrons will absorb enough energy to make a "quantum jump" to one of the higher-energy states. But within a millionth of a second or so, the electron jumps down to a lower level, in the process emitting a photon with an energy equal to the difference in the energies of the two levels. In one or more jumps, the electron returns to the ground state, emitting a photon at each jump.

Four of the possible jumps—those from level three to two, four to two, five to two, and six to two—emit photons of visible light, whereas all other jumps emit photons in the infrared and ultraviolet. All these radiations had been observed by spectrometrists long before Bohr's model accounted for them in precise mathematical detail. Other atoms, with more electrons, have different energy levels and different photons. Neon emits many visible photons, predominantly in the red. Electrical signs are long cathode-ray tubes filled with neon and other gases to give a variety of colors. The same photons observed in gases on earth are found in the light from the sun and stars, telling us that matter is the same everywhere in the universe.

Quantum Mechanics

As successful as the Bohr model was, it was still a patched-up piece of work. It explained the hydrogen atom brilliantly, but couldn't handle helium, which has two electrons. The problem is that two electrons, while being attracted to the nucleus, also exert repulsive forces on each other. This greatly complicates their motion around the nucleus, as calculated by Newtonian mechanics, making it impossible to impose a consistent quantum condition. An entirely new theory was needed, one that could handle atoms with more than one electron. This theory was the so-called new quantum mechanics, independently formulated by Werner Heisenberg in 1925 and Erwin Schrödinger in 1926.

In the new quantum mechanics, the Newtonian-like orbits disappear altogether, and, instead of describing an electron in terms of changes in its position and speed, it specifies its state. For example, the electron in the ground state of hydrogen is no longer pictured as a localized object moving from point to point around the nucleus, but as a nonlocal entity that doesn't have the property of being here or there. In classical physics, waves are nonlocal entities, so one way to describe the nonlocal nature of the electron is as a wave. The famous Schrödinger equation does this, allowing one to calculate the different waves corresponding to the various one-electron states in hydrogen, the two-electron states in helium, and so. Exact solutions exist for the one-electron case, and approximate solutions can be found for the waves in multiple-electron atoms. From its wave, various properties of a state can be calculated, such as its energy and the probability of finding the electron at different points around the nucleus. Fig. 3-4 uses shades of grey to represent the probability for an electron in the ground state of hydrogen. The darker a region is, the greater the probability of finding the electron there. Surprisingly, the electron is most likely to be found near the nucleus, unlike its orbital location in the Bohr model.

Figure 3-4 The probability of an electron in the ground state of hydrogen being in any region around the nucleus. The darker a region is, the greater is the probability of finding the electron there.

Indeed, the electron isn't orbiting at all in the Schrödinger ground state, but exists with a distribution of different speeds and positions. The average distance of the electron from the nucleus is the same as the radius of the old Bohr orbit, and the average speed of the electron is the same as the speed of the electron in the old Bohr orbit. That is, the Bohr model turns out to correspond, roughly speaking, to the average behavior of an electron in the new quantum theory. In both theories, the size of the hydrogen atom is determined by a few universal constants.

Quantum mechanics replaces particles moving along trajectories determined by initial conditions with states determined by discrete quantum numbers. From these states the probability of a particle being here or there can be calculated, but historical determinism is lost. In its place we have an absolute determinism, in the sense that the theory prescribes the states. This has bothered many great physicists because it seems to violate Mach's continuity principle (Chapter 2). Einstein couldn't accept that the fundamental laws of nature are probabilistic, stating that God doesn't play dice. But, as my artless childhood musings indicated, there are philosophical paradoxes as well with trying to picture the inside mechanism of an atom in classical terms. In his famous quantum mechanics textbook, Paul Dirac put it this way:

> The necessity to depart from classical ideas when one wishes to account for the ultimate structure of matter may be seen, not only from experimentally established facts, but also from general philosophical grounds. In classical explanations of the constitution of matter, one would assume it to be made

up of a large number of small constituent parts and one would postulate laws for the behaviour of these parts, from which the laws of the matter in bulk could be deduced. This would not complete the explanation, however, since the question of the structure and stability of the constituent parts is left untouched. To go into this question, it becomes necessary to postulate that each constituent part is itself made up of smaller parts, in terms of which its behavior is to be explained. There is clearly no end to this procedure, so that one can never arrive at the ultimate structure of matter on these lines. So long as *big* and *small* are merely relative concepts, it is no help to explain the big in terms of the small. It is therefore necessary to modify classical ideas in such a way as to give an absolute meaning to size. (1958)

There is a graininess to nature, and Planck's constant h measures the size of the grains. In particular, h determines the absolute size of atoms, and of everything that's made of atoms. At the level of individual electrons and photons, quantum mechanics describes behavior that is discontinuous with our intuitive understanding of how particles and waves behave. At the level of whole atoms and molecules, the mathematics of quantum theory blends continuously into the mathematics of Newtonian theory. Thus, quantum mechanics is logically continuous and psychologically discontinuous with classical physics.

The Uncertainty Principle

I sometimes like to tease my educator friends by reminding them that they'll never get the lowest 20 percent of their students out of the bottom quintile. This is a warning about setting unrealistic goals in the form of a reminder about an impossibility that's inherent in the concept of a distribution. The Heisenberg uncertainty principle is also a reminder about an impossibility that's inherent in the concept of a quantum distribution, although it's not a trivial one. It's a reminder in the sense that it isn't a new principle; it's a direct consequence of more basic quantum concepts, such as the particle nature of waves and the wave nature of particles.

Unfortunately, many people, who know nothing much about quantum mechanics, have heard of the uncertainty principle. The popularity of this rather arcane bit of quantum theory stems in part from the alarming term "uncertainty" and the support this gives to the common belief that quantum mechanics has introduced uncertainty and subjectivity into science. This "big lie" has entirely overshadowed the truth that quantum mechanics nails down the properties of the atoms, an absolutely incredible achievement.

One statement of the uncertainty principle is that it's impossible to exactly measure both the position and speed of an electron at the same instant. As a consequence, the exact initial conditions of an electron are unknowable, which undermines the foundations of Newtonian mechanics. This was big news when it was still thought that Newtonian determinism was essential to science. We now realize that historical determinism is good, but absolute determinism is better.

Stating the uncertainty principle in terms of possible measurements isn't meant to introduce a subjective element into the discussion. An alternative statement, which avoids this possible misinterpretation, is that the distributions of positions and speeds in a quantum state can't both be vanishingly small. For example, we know that any measurement of the position of the electron in a hydrogen atom will find it somewhere within a few billionths of an inch of the nucleus. Therefore, the uncertainty in the electron's position is a few billionths of an inch. Likewise, any measurement of the speed of the electron will find it between zero and 5,000 kilometers per second. Therefore, the uncertainty in the electron's speed is 5,000 kilometers per second. The two electrons in a helium atom have a smaller uncertainty in position—they're closer to the nucleus—but a correspondingly larger uncertainty in speeds.

The Heisenberg uncertainty principle could just as well be called the Heisenberg distribution principle. Physicists often opt to give colorful and provocative names to their concepts in order to interest the public. Unfortunately, this creates a lot of misunderstanding for those who confuse the name of a thing with the thing itself.

Social constructivists, such as Stanley Aronowitz (1988), who use the uncertainty in quantum mechanics to attack the objectivity of science, have no understanding of the subject. Just as "work" in physics doesn't mean "to earn a living," so "uncertainty" in quantum mechanics doesn't mean "doubt." Ascribing the meaning of a word from one context to its meaning in another is the sort of confused thinking that education was designed to eliminate. Unfortunately, it's all too prevalent among the educators themselves (Gross and Levitt, 1994).

In a recent address, Hans Bethe put it this way:

Many people believe that [Heisenberg's] uncertainty principle has made everything uncertain. It has done the exact opposite. Without quantum mechanics and the related uncertainty principle there couldn't exist any atoms and there couldn't be any certainty in the behavior of matter whatever. So it is really the certainty principle. (Goodwin, 1995)

Quantum Mysteries

This isn't to say that there isn't some spooky stuff in quantum mechanics. The mysteries and paradoxes of quantum mechanics have fascinated scientists for seventy years, since, in addition to being indeterminate, quantum mechanics is also nonlocal. It predicts situations in which the result of one measurement depends on the result of a second measurement which doesn't affect the first in any way. In a very famous paper, Albert Einstein, Boris Podolsky, and Nathan

Rosen (1935) argued that quantum mechanics couldn't be a complete theory because "no respectable definition of reality could be expected to permit" such nonlocal behavior. Attempts have been made to reformulate the theory in a more classically deterministic form, but at the expense of even greater complexities and nonlocalizations (Bohm, 1952). In 1964, J. S. Bell showed, on quite general grounds, that the predictions of quantum mechanics are incompatible with any local deterministic theory, that is, with any theory in which the apparent statistical behavior of quanta results from some inner machinery (hidden variables) that behaves classically (Bell, 1964). In the past thirty years, many stringent experiments, based on Bell's theorem, have been conducted with both electrons and photons to test quantum mechanical predictions, and all have been confirmed (Glanz, 1995; Kwiat et al., 1995).

All quantum mysteries come from the ability to create states of two electrons or photons that are spread out over large distances. It's one thing to have electrons spread out around an atom; it's quite another to have them spread out around a room. To give a flavor of how spooky this can be, we'll consider a classical interference experiment. Such experiments were done with light long before quantum mechanics, and they continue to be used, in one form or another, in many branches of physics.[5] Electrons exhibit exactly the same interference phenomena as photons do, but experiments with light are easier to conduct and to describe.

In the simplest case, a beam of laser light falls on a sheet of glass with a thin silver coating that reflects half of the light, and allows the other half to pass through the glass (Fig. 3-5 (a)). In effect, the original beam is split into two equal beams that now travel along different paths. Using two fully reflecting mirrors and a second beam-splitting mirror, the two beams are brought back together and the recombined beam falls on a screen or photographic plate.

Because the two halves of the beam have traveled different distances, their waves will combine differently at different points on the plate. At some points, such as A in Fig. 3-5 (b), the waves are in step and add together to give a bright spot on the plate. At other points, such as B in Fig. 3-5 (b), the waves are out of step—a crest of one always canceling a trough of the other—and no spot appears. The recombined waves are said to interfere with each other, producing an interference pattern of light and dark regions.

All of this is in accord with classical wave theory. But now quantum theory tells us that light is composed of photons, so we can make the initial laser beam so faint that only one photon at a time falls on the partially silvered mirror. Only one photon at a time reaches the photographic plate, but over time a pattern emerges from the accumulation of many photons. With only one photon in the system at a time, there should be no interference; some photons should reach point B. But none do. The same interference pattern emerges from this experiment as from one with an intense initial beam of light.[6] How can this be?

One possibility is that the partially silvered mirror splits each photon in two, and the two half-photons interfere with each other. This has been carefully tested, using the photoelectric effect, and it's definitely not the case (Clauser,

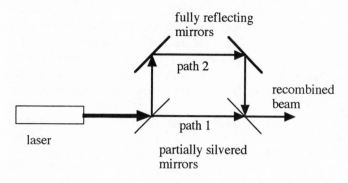

a)

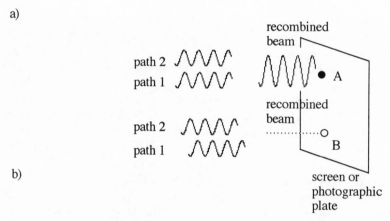

b)

Figure 3-5 a) Interference experiment in which a laser beam is split by the first partially silvered mirror and the two split beams are recombined by the second partially silvered mirror. b) Waves from the two beams are in step at point A on a screen and so add to give a bright spot. Waves from the two beams are out of step at point B on the screen and so they cancel each other out.

1974). Alternatively, some photons might be reflected by the mirror and travel along path 1, and others are transmitted and travel along path 2. In this case, however, each photon arriving at the plate would have followed either path 1 or path 2, and there would be no interference pattern. To account for the observed pattern that is built up from a succession of individual photons, one must assume that the mirror creates a nonlocalized state in which each photon travels along both paths.

"That's impossible," you say. "The photon must be one place or the other. It's just that we don't know which, right? It's that uncertainty thing again."

To test this idea, let's place an automatic gate in each path. These gates are programmed to randomly block one path or the other. That is, for a few microseconds path 1 is blocked, and no photon moving along it can reach the plate; then the block is removed, and instantly path 2 is blocked. The periods of blocking vary randomly. What will happen to the pattern?

At all times, one path is open and one path is closed. Thus if each photon is in one path or the other, the closed gate will sometimes block a photon and sometimes not. When it does block a photon, no light reaches the plate; but when it doesn't, the photon in the unblocked path reaches the plate. Half the initial photons should reach the plate along unblocked paths and the pattern should be the same as it would be without the gates, only fainter (since fewer photons get through).

But this isn't what happens. Blocking one path or the other completely destroys the interference pattern. This is the mystery of nonlocalization; the behavior of a photon is affected by a gate in a path it didn't take. This idea seems paradoxical, because it's completely contrary to our intuitive ideas of when one thing can affect another. But, from another point of view, the experiment with the gates is physically different from the experiment without the gates, and there is no paradox in two different experiments having two different results. These two physically different arrangements of mirrors and gates create two different quantum states, one that is extended over both paths and one that isn't.

This is the standard interpretation of quantum mechanics, as developed by Bohr and his colleagues in Copenhagen between 1925 and 1935. It avoids the paradox of nonlocality by not making statements about what isn't observed. Since quantum mechanics correctly predicts the results of any particular laboratory experiment, the theory is complete; it says all that can be said about the nature of atomic phenomena (Bohr, 1958; Stapp, 1972). Puzzlements arise only when, in the course of one experiment, one asks about the results of a complementary experiment. For example, one is puzzled by the notion that a single photon must be simultaneously traveling along two spatially separated paths in an interference experiment. This is because determining which path the photon takes is a different experiment, complementary to the original one. Putting photon detectors in each path will find that each photon is, in fact, in one path or the other, but the interference pattern will then vanish. These puzzlements continue to amaze and fascinate (Peres, 1978; Jordan, 1994; Mermin, 1994), but they don't undermine the objective nature of quantum mechanics.

In Mach's positivism, theory is just an economical device for relating a less familiar phenomenon, such as the motion of a planet around the sun, to a common everyday experience, such as the falling of an apple. To accommodate quantum theory and relativity, logical positivism freed theory from the requirement that it ultimately connect to intuitive knowledge and promoted it to an independent status comparable to experience. As described in Chapter 2, science is the connection between theory and experience.

Chaos

Quantum mechanics narrowly construes indeterminism to mean only that the theory predicts not where a particle will be at a certain time, but the probabilities of it being in various places. This is sometimes hailed as freeing us from the clutches of Laplacian determinism and explaining free will. But confusing physics and theology isn't beneficial to either.

Recent thinking on chaos and determinism has turned the issues upside down. We now know that most Newtonian systems, even very simple ones, aren't predictable. There's nothing new in this. Newton showed that *one* planet would move in a perfect ellipse about the sun, but neither he, nor his successors, ever found an exact expression for the motion of *two* planets moving about the sun. If the planets are far enough apart, their mutual gravitational attraction causes only a slight departure from perfect elliptical motion, and this departure can be calculated to good approximation. But if the planets are close enough to affect each other strongly, the approximations are limited in how far into the future they can predict the motion. Thus, it has been known since the eighteenth century that Newtonian mechanics gets very complicated as one deals with any but the simplest systems. What is new today is the word *chaos* to characterize complex behavior and computers with which to study it.

We now know that unrestricted predictability is characteristic of only a very few idealized systems—the ones we teach in physics courses—and that most Newtonian systems, including the solar system itself, are chaotic. A change of only a millimeter in the assumed position of Neptune, causes uncontrollably large changes in the calculated positions of the other planets a hundred million years in the future (Sussman and Wisdom, 1992). The solar system is deterministic because future positions can, in principle, be calculated from past positions, and unpredictable because these calculations become increasingly unreliable the further into the future they are carried.

A Newtonian atom would be even more chaotic than a Newtonian solar system. The Newtonian motion of one electron orbiting a hydrogen nucleus is an ellipse, just as is the motion of one planet orbiting the sun. But the Newtonian motion of two electrons orbiting a helium nucleus is chaotic. This is because there is a strong electrical repulsion between the electrons. The behavior of such a system can be studied by programming a microcomputer to calculate the Newtonian motion of two electrons orbiting a positively charged nucleus (Gould and Tobochnik, 1996; Yamamoto and Kaneko, 1993.). The program starts the electrons in circular orbits about the nucleus, but they quickly move into more complex orbits until one of them is ejected from the atom and the other settles into a one-electron orbit around the nucleus. But when this occurs, or even which electron stays and which leaves, is too sensitive to initial speeds and positions to be predictable. That is, a stable Newtonian helium atom couldn't exist.

There is no chaos in quantum mechanics because initial conditions, in the classical sense, don't exist. The two electrons in a real helium atom occupy a stable ground state with a well-defined energy. A helium atom is indeterministic because its electrons can't be thought of as occupying successive points at successive times, and highly predictable because all helium atoms have exactly the same ground state and sequence of excited states. In short, because quantum systems aren't chaotic, they are more predictable than most Newtonian systems (Jensen, 1992).

Conclusion

My childhood question about how structureless atoms hook together was answered in graduate school, where I learned that it had something to do with nonlocalized electrons being in two places at the same time. A mystery explaining a mystery. What's important is that by injecting uncertainty at the level of the electron, quantum mechanics gives us great constancy and certainty at the level of atoms, molecules, and everything larger.

Hydrogen and oxygen atoms hook together (that is, share electrons) in a precise way, so that one water molecule is the same as another. The same is true for all other molecules, which explains why we live in a world of differentiated substances. All chemical reactions, from burning charcoal to DNA replication, proceed according to their own strict rules and regulations. This absolute determinism is a direct result of the regularities of historically indeterministic quantum mechanics. Were the world Newtonian to the core, it would be even more chaotic than it is. Perhaps we live in the least chaotic of all possible worlds.

By explaining the fundamental regularities of nature, quantum mechanics strengthens and unifies the entire scientific enterprise. Its own foundations have been tested and retested exhaustively for the last seventy years, and it has passed every test anyone has ever thought to put it to.[7] Theoretical calculations of the energy levels of a hydrogen atom agree with experimental measurements to six decimal places, a spectacular demonstration of the objective validity of scientific knowledge.

Unfortunately, these achievements aren't understood by most academics. Increasingly, students and student teachers are being taught one brand or another of postmodern philosophy that rejects all claims of objective knowledge. This "rejectionism" started with the literary critics, who projected the nonobjectivity of their own work onto everyone else's. It then spread to sociology, history, education, and science education, where now much of the learned discourse is about learned discourse, entirely disconnected from experience and practice. One obvious explanation of this is that the social sciences are inherently unscientific, so that these academics have nothing better to do with their time. If this is true, then what special right do historians and sociologists have to speak on any issues affecting political and social life? If it's not true, what are the repeatable phenomena and strong predictive theories in the social sciences? These questions are the subject of Chapter 4, which reaches conclusions rather different from those that I had anticipated before undertaking this study.

4

Science in the Social Sciences

The social sciences encompass a broad range of academic subjects, including history, psychology, sociology, social anthropology, social psychology, education, and so on. Even a specialty such as social psychology gets divided into numerous subspecialties, each with many models and theories, resulting in almost fractal fragmentation. What is one to make of all of this? Can the social sciences be considered sciences in any sense comparable to the natural sciences? The question is important because it's the social scientists who tell us how to fight poverty, educate our children, and formulate foreign policy. And since many social scientists are leading the postmodernist attack on the natural sciences (Gross and Levitt, 1994), it's not inappropriate for me to examine the foundations of their discipline.

The two essential ingredients of science are repeatable phenomena and strong inferential theories, because science is the struggle for universal consensus. The evidence for any claim must be readily available to anyone who would question the claim. Science always operates in the present, so every claim must be verifiable by evidence currently available. Theoretical constructs must be strong enough to cover a range of specific instances and precise enough to lead to definite conclusions.

Historical Sciences

At first thought, the repeatability criterion would seem to eliminate the historical sciences—history, archeology, and paleontology—from the halls of science. One can't go back and rerun the Revolutionary War to see whether things would have worked out differently had the British Parliament not passed the Stamp Act. All that's known about the past are the traces left in the present.

It's these traces—fossils, shards, letters—that are the phenomena of historical research. Repeatability doesn't apply to the historical event itself, but to the processes of acquiring and analyzing the evidence on which conclusions about the past are made. This is clearly one step removed from the methods of the natural sciences, but as long as investigators can search for their own bones and pots, or can examine those found by others, the critical reexamining process of science remains intact. Because of the priceless value of many important specimens and artifacts, this reexamining process is necessarily limited, which must certainly slow progress in these fields.

Controversies in human history arise not from any inherent lack of objectivity in its methods, but from the catastrophic nature of its subject matter. The most important facts about the assassination of President Kennedy are not in dispute: he was shot to death in Dallas, Texas, on November 22, 1963. What is in dispute is a relatively minor matter—whether Lee Harvey Oswald was the assassin, acting alone or as part of a conspiracy—which has catastrophic consequences for the interpretation of the event. In effect, the world splits into parallel realities on the basis of this dispute. Evidence that's compelling enough to achieve consensus in other matters isn't in this one.

More generally, historical explanations are hindered by the lack of a strong inferential theory that could allow interpretations of one circumstance—say the French Revolution—to be confirmed by another—say the Russian Revolution. A strong theory would, as the theory of an oscillating rod (Chapter 2), identify the relevant factors in a situation and assess their relative importance. Economics is certainly a major factor in human affairs, but it is so strongly affected by other factors that predictions based on it are problematic.

Nevertheless, insofar as the historical sciences are about interpreting existing documents and artifacts, and insofar as these artifacts can be freely examined by scholars, the scientific requirement of repeatability is met. When a major scandal erupted over alleged misquotations in David Abraham's *The Collapse of the Weimar Republic* (1981), scholars could review the original documents themselves to judge the seriousness of the discrepancies (Campbell, 1984). Postmodern historians aren't impressed with this concern with the factual evidence, dismissing it as mere "facticity" when it interferes with their own notions of historical truth. Unfortunately, this loss of faith in objectivity is now quite respectable in high academic circles (Novick, 1988).

History, as an academic discipline, is more recent than physics. Its Galileo was Leopold von Ranke (1795–1886), the founder of modern historical research. His simple desire to show how things actually were (*wie es eigentlich gewesen*) became the motto of generations of historians. He tried, as much as possible, to keep his own strong opinions out of his work, for which he was much criticized in his time. He was the first to favor primary sources (letters and documents) over memoirs and chronicles, and to weigh the evidence of authorities according to their proximity to the events about which they wrote.

> He was the greatest historical writer of modern times, not only because he founded the scientific study of materials and possessed in an unrivalled degree the judicial temper, but because his powers of work and length of life enabled him to produce a larger number of first-rate works than any other

member of the craft. It was the Goethe of history who made German scholarship supreme in Europe, and he remains the master of us all. (Gooch, 1913)

Implicit in all the historical sciences are two overarching principles that are so obvious that they are seldom explicitly mentioned. One relates to time, and is as general as any in physics:

> *The Chronology Principle.* For any two events A and B, it must be the case that either A occurred before B, A and B occurred simultaneously, or A occurred after B. Moreover, these relations satisfy the same properties as the mathematical symbols $<$, $=$, and $>$. Among these is the transitive property, which says that if A occurred before B (A $<$ B), and B occurred before C (B $<$ C), then A occurred before C (A $<$ C).

This is the basic organizing principle of all the historical sciences, enabling historians to arrange events in order. We take dates so much for granted that we forget what a great accomplishment the establishment of absolute chronologies is. It was motivated by the Christian belief that the Old and New Testaments give a true and continuous history of the world from the creation of Adam and Eve to the coming of Christ.[1] To convince nonbelievers of this history, it was necessary to synchronize biblical with nonbiblical events by establishing an absolute chronology. In the third century, Sextus Julius Africanus produced the first such chronology, which started from creation and put the birth of Christ in the year 5500. The modern method of dating from the birth of Christ was introduced in the sixth century by the Roman monk Dionysius Exiguus. Over the centuries many eminent scholars, including Newton, have contributed to this task, so that we know that Aristotle died in 322 B.C. and that the age of dinosaurs ended sixty-five million years ago, plus or minus 100,000 years.

Another historical principle relates to matter:

> *The Continuity Principle.* Objects do not come into existence spontaneously, but are formed, either naturally or artificially, from other objects. Any object existing now has existed continuously since it was first formed.

This principle is assumed whenever an object is traced back in time to some original formation. Thus a fossil of a fish is believed to be formed from a real fish that lived millions of years ago, by the continuous replacement of fish atoms by rock atoms. The oldest existing version of the *Iliad* is believed to be a copy of earlier versions, forming an unbroken chain stretching back to the first written version In both cases, the processes of preservation are not perfect, but they provide us with the only evidence we have of ancient fish and literature.

It must again be stressed that these principles, as reasonable as they sound, can't be proven and aren't the only principles possible.[2] One can replace the continuity principle with the biblical principle that says that the Old and New

Testaments are literally true. Then, fossils aren't interpreted as the remains of ancient plants and animals that lived millions of years ago, but as objects especially created by God to indicate ancient life-forms that never existed, just as He created Adam and Eve with navels to indicate a natural birth that hadn't occurred (Gardner, 1981). This tortured logic suspends the continuity principle to save Genesis, just as Velikovsky suspended the laws of mechanics to save Joshua (1950). Nonreligious suspensions of the chronological principle are found in stories of time travel, from H. G. Wells's *The Time Machine* to the movie *Back to the Future*. In *Einstein's Dreams,* Lightman (1993) imagines no fewer than thirty fabulous ways that time might behave. And *Hyperspace* (Kaku, 1994) speculates about the possibility of connecting parallel universes, which would wreak havoc with both the continuity and chronology principles.

None of these alternative principles rival the chronology and continuity principles in generality, logical consistency, or conformation with experience. All the mainstream historical sciences rely on the chronology and continuity principles, and so have a common philosophical and methodological foundation. All alternative versions of history use idiosyncratic principles, often unstated, that arise from a particular interpretation of history. One example is the claim of some Afrocentrist scholars that Aristotle stole his ideas from the library of Alexandria, even though the library was established after his death (Lefkowitz, 1996).

It's sometimes argued that general principles, such as the chronology principle, can't be disproved because the historian can always rearrange the evidence to conform to them. An often cited example are fossil beds in which older fossils are found on top of younger ones. To a geologist, this indicates that the ground has been uplifted and thrust over, pushing an older bed over a younger one. To a creationist, it indicates flaws in the principles of geology. The creationist would have a valid point if the geologist had no evidence for the uplifting other than the inverted time order. But she does: the same beds, with the same fossils, are found a few miles away in the expected order. It's true that an occasional anachronism will be dismissed as a hoax or dating error, but the chronology principle would be disproved if there were convincing historical evidence that George Bernard Shaw helped Shakespeare write his plays.

Discussions of history and history education generally overlook the solid theoretical and factual foundation of history. The recently published *National Standards for United States and World History* does put an emphasis on chronological thinking and rigorous scholarship, but the sheer bulk of the material to be covered, and *Standards'* persistent judging of the past in terms of the present, makes it unlikely that students will learn how things actually were (History Standards, 1995; McDougall, 1995). Without a disinterested study of the facts of history, and how they are discovered, the Standards will just replace one set of myths with another.

The foundations of history are as ellusive as those of science; they aren't intuitive and they won't be learned if they aren't taught. Few students have any sense of how historical events relate to one another, either causally or chronologically. Many don't yet appreciate the causal connections in their own lives.

The whole question of causation in history is problematic anyway, since there is no widely accepted theory of human behavior. Whether one considers the African slave trade or the Abolition movement the extraordinary event depends on your particular view of human nature.

There is probably a sequence of steps that students must take to develop historical understanding, just as there is a sequence for developing scientific understanding. Perhaps memorizing the Presidents of the United States in chronological order should be the starting point of any study of American history, or history in general. This isn't because the Presidents were necessarily the movers of history—though often they were—but because they serve as convenient markers of the timeline of history. Such a timeline teaches about chronology while providing a memory palace in which to place all other events of the last two centuries. The Chinese Revolution occurred during the Presidency of William Taft, and so it came after the U.S. Civil War (Lincoln) and before World War I (Wilson). Although this sort of memorization is difficult for adults (I had to look up most of this), first- and second-graders have great capacity for this sort of learning. Seven-year-old boys delight in memorizing the starting players on dozens of baseball teams over scores of years. In Greece, students memorize the emperors of the Byzantine Empire, from Constantine the Great (305) to the fall of Constantinople (1453).

Academic historians are highly specialized. With the demise of survey courses in Western Civilization, they are seldom called upon to teach outside the country or century they wrote their theses on. By its nature, scholarly research investigates what's controversial and uncertain about past events. There's no doubt that Newton undertook the writing of the *Principia* after he was visited by Halley. And there's no doubt they talked about calculations pertaining to the elliptical motion of planets moving under the influence of an inverse-square force. But whether Halley asked what the curve would be if the force was an inverse-square, or what the force would be if the curve was an ellipse, is the level of detail that engages research scholars (Cohen, 1971).

Scholarly research, as scientific research, is necessarily about unsettled, often arcane, issues. This is as it should be. The trouble is that too often scholarly controversies and uncertainties are presented to students who don't know the basic facts of history. Naturally, then, they come to think that all history is just a construction that can be interpreted howsoever they please. *National History Standards,* as *National Science Education Standards,* want students to be encouraged to think critically, without emphasizing that training and knowledge are necessary if opinion and prejudice aren't to be passed off as analysis.

Once at a meeting of my college's core curriculum committee, I waxed nostalgic for the long-gone survey courses in Western Civilization. This got the expected rejoinder, by a male professor, that these courses had excluded women and non-Western civilizations. "Perhaps," I said, "but students today don't know whether Moses lived before or after Mohammed. They don't know the basic facts of history." This got the very unexpected acknowledgment from a women professor that perhaps students can't know who was excluded from history if they don't know any history at all.

Categorization

Chronology and continuity are strong organizing principles for classifying and ordering historical events. There are no comparable principles for categorization in the sociological disciplines, yet categorization remains the paramount methodology in the nonhistorical social sciences. What is the philosophical justification for this?

Categorization seems to be a natural mental process, closely linked to language. It's neither possible nor functional to have a different word for every object and activity in the world, so words come to represent classes of things. "Walking" refers not just to the walk I took today, and not just to all my walks, but to the walks of all human beings, and by extension, to similar activities of other legged animals. "Bird" refers not just to a particular robin in my back yard, and not just to all robins, but to a large class of feathered animals. The question that has plagued philosophy for years is whether such categorizations are an imposition of the mind on an undifferentiated reality, or whether they are names given to preexisting differences in nature. The answer, I believe, is an unequivocal "it depends."

In itself, categorization is a weak and often arbitrary mental process. Eight-year-olds may collect and arrange any manner of things—dolls, stamps, baseball cards, rocks—in any manner of ways. Whether any particular classification scheme makes sense depends on one's understanding of the things being classified. If there's no theoretical basis for the classification scheme used, the categories may indeed be figments of the classifier's imagination, such as a child arranging her dolls according to how good they've been that day.

On the other hand, some similarities are so apparent, as among different species of birds, for instance, that the grouping "birds" seems natural and objective. The classification of animals, which dates back to Aristotle, was for thousands of years based on similarities and differences that are apparent to the senses. But what similarities and what differences? In ancient times, whales were classified with fishes, because they all lived in water. Later, however, whales and dolphins were seen to resemble mammals anatomically, and this method of classification was adopted by Carolus Linnaeus in the eighteenth century.

However, without a theory, there's no rational way to decide between a classification based on habitat and one based on anatomy. The latter may seem more scientific, especially as it often requires the detailed comparisons of teeth and bones, but it doesn't yield unambiguous results. In the nineteenth century, sea squirts were classified as mollusks, because they live attached to the sea floor. Their free-swimming larva, however, have a notochord, which places the sea squirt, along with us, in the phylum chordata.

The principle of evolution extends the principles of chronology and continuity to biology, providing a theoretical basis for taxonomy that is very different from what had preceded it. "All true classification is genealogical; that community of descent is the hidden bond which naturalists have been unconsciously seeking, and not some unknown plan of creation, or the enunciation of general propositions, and the mere putting together and separating objects more or less alike" (Darwin, 1859). The genealogy of all living organisms extends back to the first living creature, some 3.5 billion years ago, a startling assertion that offended Christian doctrine in its time, and continues to offend creationists today.

We can contrast the state of affairs in biology to that in the social sciences by examining two of the classification schemes developed by educators to categorize students' learning styles. According to these educators:

> Learning style is a biologically and developmentally imposed set of characteristics that make the same teaching method wonderful for some and terrible for others. . . . Many well-conducted, experimental studies demonstrate how well the same youngsters learn when they are taught correctly (for them) and how poorly they learn when they are taught through methods that do not complement their styles. (Dunn and Griggs, 1988)

This concept is very seductive because it fits into the current educational ideology that all children can learn—if only the teachers would do their job right. Part of their job is to identify their students' learning styles and take appropriate actions to accommodate them. For example, in the classification scheme of Dunn and Dunn (1978), students are divided into those who prefer quiet or noise; these are further divided into those who prefer bright or dim illumination. This gives four categories: quiet–bright, quiet–dim, loud–bright, loud–dim. Adding a temperature preference (hot–cold) doubles the number of categories to eight; adding seating preference (desk–table) doubles the number to sixteen, and so on. Dunn and Dunn actually test for twenty-three distinct preferences. If each preference were scored dichotomously (this–that), the teacher would face 8.4 million different learning styles. Actually, this system scores on a scale of five (strongly prefer this, moderately prefer this, no preference, moderately prefer that, strongly prefer that), which results in over 10,000 trillion categories, enough to give a different learning style to every child born in the next 200 million years. Try validating that system!

Very different is the 4MAT system, which classifies students along two axes according to their answers to a multiple-choice questionnaire. One axis measures the student's preference for acquiring knowledge concretely versus abstractly, and the other measures his preference for applying knowledge concretely versus abstractly (McCarthy, 1980). His scores on these axes places the student into one of the four quadrants: 1. Innovative Learners (acquire concretely, apply abstractly); 2. Analytic Learners (acquire abstractly, apply abstractly); 3. Common Sense Learners (acquire abstractly, apply concretely); 4. Dynamic Learners (acquire concretely, apply concretely). Each of these quadrants is divided again, according to whether the student prefers processing information analytically (left-brain mode) or holistically (right-brain mode). The 4MAT System trains teachers to write lesson plans that cycle through all eight learning styles and brain modes.

Such more-or-less arbitrary classification schemes abound in the social sciences. They are easy to dream up, and virtually impossible to validate (Wilkerson and White, 1988). This doesn't mean that they're totally useless. A teacher may feel more positive toward a spacy student once she realizes that he's just doing his category-one thing. She may also adopt more varied teaching methods as she realizes that most students respond well to demonstrations and visualizations. But this "usefulness" doesn't validate a learning-style system any more than deciding to act on your horoscope validates astrology.

Learning style systems are similar to much of social science research in their penchant for broad categorizations. These categorizes are clearly conventional, being an invention of the investigator, not nature. Nevertheless, some learning-style schemata make sense to some teachers, who become enthusiasts. For them, a particular schema provides a viable solution to their immediate classroom problems. Such schemata can have no claim to objective truth, and none is seriously made. As with astrology, any remark about the arbitrariness of the categories is answered with "I know, but it works for me."

These remarks aren't meant to ridicule the end product of the different learning-style systems, which in all cases is the greater use of different modalities: verbal, aural, visual, tactile, and kinesthetic (Dunn and Griggs, 1988). All students—indeed, all adults—prefer activities that involve all their senses and motor skills. I have yet to meet a student or teacher who isn't captivated by building a motor and watching it spin around. But this conclusion doesn't follow from learning-style theory, which has each student preferring a different environment and experience.

Interestingly, recent research shows that the cerebellum, long thought to be a purely motor structure of the brain, may be connected neurologically to cognitive regions of the prefrontal cortex (Middleton and Strick, 1994).[3] Though still controversial, there is growing evidence for an anatomical connection between motor skills and cognitive skills (Barinaga, 1996).

Sociological Disciplines

The academic disciplines of psychology, sociology, social psychology, and social anthropology are currently in chaos and disarray. Through a process of fragmentation and specialization, each field has divided and subdivided into ever smaller units of inquiry. In social psychology, for example, there are separate studies of attribution, motivation, social cognition, self-perception, action, attraction, attitude, prejudice, social relations, and on and on, each with as many competing theories and categorizations as there are senior investigators. Moreover, the theoretical concepts are too vague to be applied in any consistent manner, and they don't make use of work in related fields. The attitude literature makes no connection with the prejudice literature, the social-relation literature doesn't refer to the social-cognition literature, the motivation literature has little in common with the action literature (Vallacher and Nowak, 1994).

Natural scientists usually just sneer at this mess, dismissing the social sciences as nonsciences. Or else they say that human behavior is just too complex to model. But I take a different position. Not only do I believe that a science of human behavior is possible, I believe that such a science exists and stands on a permanent foundation of fundamental knowledge.

For example, we know that human beings are part of the biological order, made of the same physical, chemical, genetic, and cellular material as are all other animals. We know that our species evolved some 200,000 years ago from hominid ancestors that separated genetically from the apes' ancestors some five million years ago. We know that when this evolution occurred our ancestors were living in small roving bands of hunter-gatherers, and that until 10,000 years ago, few human beings ever saw more than a few dozen other human beings.

All this is so well known that one might assume that it's the starting point of any theory of human behavior. But the fragmented subsubspecialities of the sociological disciplines hardly refer to one another, let alone to the great body of pertinent scientific knowledge outside the speciality. Graduate and undergraduate sociology and anthropology majors have only minimal requirements in science and mathematics, mirroring the preparation of their professors. Thus by selection and training over several generations, the sociological disciplines have completely isolated themselves intellectually, philosophically, methodologically, and socially from the natural sciences.

This is a degeneracy from the positions of the nineteenth-century founders of experimental psychology, who placed their work firmly in the scientific tradition. Wilhelm Wundt, William James, and G. Stanley Hall were assistants and students of the physicist Hermann von Helmholtz, who had turned his great analytic and experimental techniques to the study of human perception.[4] The work of nineteenth-century psychophysics established the principles that relate the physical properties of sound and light to the human perception of loudness, pitch, brightness, color, and so on—principles which form the basis of today's acoustic and color technologies. In 1884, F. Holmgren found that subjects observing a flashing point of white light at the threshold of visibility, perceived the light sometimes as red and sometimes as green. This was the direct observation of individual photons striking a red or green photoreceptor in the retina, two decades before Planck and Einstein introduced the quantization of light into physics (Krauskopf and Srebro, 1965). Work in perception continues today, most notably in the area of stereovision. From the pioneering work by Bela Julesz on random-dot stereograms, has come the popular computer-generated random-dot autostereogram (Julesz, 1971; Tyler, 1983, 1994).[5]

Other psychologists, notably B. F. Skinner, developed "behaviorism," a methodology for studying human and animal behavior that was closely modeled on positivism in the natural science. It avoided inventing terms for invisible mental processes, focusing instead on the visible factors that affected visible behavior. Its theory of learning had strong predictive power when applied to routine learning tasks (Chapter 8), but it was zealously overpromoted as a radical solution for accelerating higher-level learning. Many social scientists were relieved when crude learning machines and other experimental technologies failed to live up to their exaggerated claims, because it seemed to discredit behaviorism altogether.

So, following the pioneering work of Jean Piaget, psychology switched from doing controlled experiments on isolated aspects of behavior, to studying full-blown cognitive functioning. For example, instead of observing the behavior of pigeons and mice, a cognitive psychologist might ask students how they solved a physics problem and what their understanding is of force and acceleration. As Piaget had found that children think differently from adults, cognitive researchers found that students think differently from their teachers.

These were interesting findings, perhaps, but difficult to interpret. Theoretical models were developed in education, and in many other branches of psychology and social psychology, that no longer adhered to strict positivist principles. It became acceptable to developed an arcane vocabulary for what goes on inside the human mind, totally disconnected from observable phenomena. Metaphysics returned to the social sciences, big time. Untethered concepts became the objects of an ever-more-diffuse rhetoric, leading to the fragmentation of psychology and related fields into ever-smaller subdivisions with their own journals, conferences, and research standards. These standards dropped to appallingly low levels, with every paper containing a new "model" with its own terms inscribed in circles all connected to one another by two-headed arrows. In short, words replaced reality.

Social Constructivism

It isn't surprising, perhaps, that it was within the highly fragmented sociological disciplines, with their many competing theories and categorizations, that a new doctrine emerged—social constructivism—that not only justifies this academic tribalism, but sets it philosophically above the natural sciences. Social constructivism sees all knowledge—or rather, all illusions of knowledge—as derived from social interchanges, instead of seeing social interchanges as deriving from more basic biological principles (Gergen, 1988). In this inverted theory of knowledge, science, philosophy, and the philosophy of science are all reducible to sociology and are subject to the harsh criticisms of the social constructivists, who cast doubt on all claims to objectivity. In *Higher Superstitions,* Gross and Levitt document the full extent to which this, and other relativistic and "perspectivist" doctrines have established themselves among faculty in the humanities and social sciences (1994).

One does not have to be a social constructivist to see who wins and who loses by dethroning objectivity. Nor is it particularly surprising that constructivist doctrines arose among those academics most ignorant of science. Some humanists and social scientists may have read Kuhn, but very few took more than the minimum science requirements in college. The consequences of this run very deep. A theology professor, after auditing an introductory physics course as part of a study of introductory sciences courses, commented:

> Some of the concepts were not entirely unfamiliar but they were used very differently. . . . The idea of zero or zero-ness. Unless a nonphysicist deliberately thinks about it, "0" is the absence of anything, the absolute bottom or

"start." But to the physicist, zero is actually in the middle, with plus and minus quantities on either side. (Tobias, 1993)

This worthy academic has the mathematical sophistication of the drug dealer who told me he had learned in algebra that "zero has no credibility."

The problem with the sociological disciplines, then, is not that human behavior is unsuited to scientific analysis, but that the professorate in these fields is largely ignorant of the most elementary principles of science and scientific inquiry. The result is a rout of objectivity and standards, and the institutionalization of ignorance.

Data and Analysis

Quantitative research in the social sciences often involves the collecting of data on groups of people through surveys and tests of various sorts. This work originated from the need of the tax collector to determine the status—or statistics—of the population: the number of citizens, slaves, births, deaths, and so on. About 6 B.C., "Caesar Augustus issued a decree that a census should be made of the whole inhabited world" (Luke 2:1), and governments have been doing it ever since. Such descriptive statistics comprises a large part of the work that social scientists do, and as such it's purely empirical.

To go beyond mere census taking, it's necessary to make some inferences on how one variable (say, income) depends on another (say, education). This is usually done through some statistical analysis of the data that tests for a correlation between the variables. Fig. 4–1 shows the data on income and education used in *The Bell Curve* (Herrnstein and Murray, 1994) to show what typical data in the sociological disciplines look like. The dots represent the income and years of education of individuals with incomes under $100,000, randomly selected from a much larger database. There's always an enormous amount of scatter in such plots, even of variables such as education and income for which a strong correlation is suspected. This is because individual behavior is affected by many factors of more or less equal importance. Many factors also affect the period of the oscillating rod discussed in Chapter 2, but they are of different importance. The length of the rod and the position of the hole account for 98 percent of the rod's period, whereas education accounts for only 10 percent of an individual's income. Moreover, there is a strong theory for the oscillating rod that tells us just how the less important factors enter into the equation, so they can be studied one by one. There is no such strong theory in the social sciences.

Social science data are neither as repeatable or as reliable as physical science data. Although a survey of income and education done in 1989 can be repeated in 1997, it would involve a different group of individuals living under somewhat different social and economic conditions. And the reliability of data based on individual responses to a questionaire is limited by people's truthfulness. If people lie randomly, the effect is to increase the scatter of the data, washing out any suspected trend. If only the marginally employed are inclined to understate their income, the data may overstate the effect of education on income. There

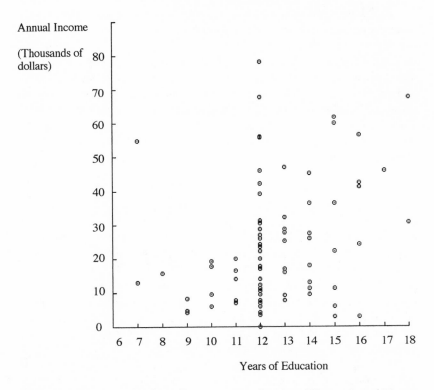

Figure 4-1 Representative data of income versus education. (Herrnstein and Murray, 1994).

are, in addition, sampling and statistical problems. How representative is a sample population of the population as a whole? This can be a problem with studies done on small populations of college students, and even with large national data sets, when looking at rare events. An example of this is the small number of cases of childhood leukemia in the Swedish study of half a million people living near high-voltage power lines (Chapter 3).

Statistical analysis is most beneficial when it is used to show that a suspected correlation, say between proximity to a power line and leukemia, is just a random fluctuation. That is, it can tell you that less is there than meets the eye, but seldom does it find more. Before reading further, take a few minutes to study Fig. 4-1. What conclusions can you draw from it? The human eye-brain system is highly developed for recognizing visual patterns. Its bias is sometimes to see things that aren't there, like faces in the cloud, but it seldom misses a real face.

Your first impression of Fig. 4-1 was probably that the points are so widely scattered that they show no consistent trend. Looking closer, you may have noticed that there are more high incomes on the right than on the left, indicating a small positive effect of education on income. Can more be said? Does the data support the often-made assertion that each year of schooling is worth so-many dollars of income?

The most frequently used method for analyzing data like that in Fig. 4-1 is linear regression, a mathematical procedure for finding the straight line that best fits a data set. It's taught in all statistics courses and has been routinely used by social scientists for over a century. With the right software package, or statistical calculator, anyone can learn to do such an analysis is a matter of minutes. But linear regression is applicable only in situations in which there is a theoretical reason to believe that the two variables are linearly related (Kohler, 1988). We have such theories in physics, and students in the physics laboratory are routinely asked to fit a straight line to their data. If the data are plotted on graph paper, a "best-fit" line can, with reasonable accuracy, be drawn with ruler and pencil. This low-tech approach works when the data do, in fact, lie more or less on a straight line.

The data in Fig. 4-1 do not immediately suggest that there is a straight-line relationship between income and education, but if you believe that there is, you can use the ruler-and-pencil approach to draw a straight line that "best fits" the data. Or, if you're highly educated, you can use a computer to do it for you. Fig. 4-2 shows the computer-calculated regression line that splits the differences between the points according to a precise mathematical protocol. This line does rise by a fixed dollar amount—$2,600—with each year of education.[6]

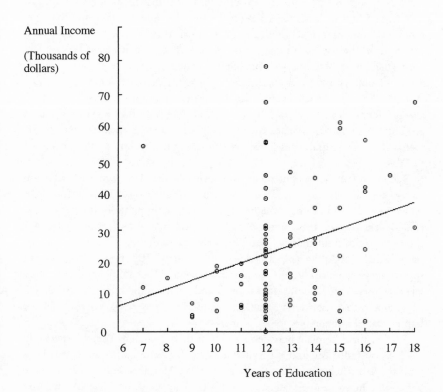

Figure 4-2 The income versus education data for Fig, 4-1, with the best-fit straight line. The slope of the line is $2,600 per year of education.

From such analyses come the statement that each year of education is worth so-many dollars of additional income.

But the conclusion that there is a fixed increase in income for each year of education comes from an analysis that is valid only if there is a theoretical reason, independent of the data, for income to be linearly related to education.[7] There is reason to believe that the number of miles I drive a month is linearly related to the number of gallons of gasoline I buy. A linear regression of my monthly mileage and gasoline purchases would yield a number—miles traveled per gallon of gasoline—that has a bearing on the condition of my automobile. Likewise Planck determined the numerical value of his constant h by fitting his theoretical formula for black-body radiation to the data (Chapter 3). Statistical analysis helps to determine certain numbers, or parameters, in a theory, but it doesn't determine the theory. Since there's no theoretical reason to suppose that income is linearly related to education, the number generated by a linear analysis has little significance.

Theory must precede analysis, since the appropriate analysis is determined by the theory. For example, consider the certification theory, which says that income is related to certification, not to the number of years of schooling. The theory is based on the prior knowledge that failure to get a high school diploma cuts a person out from almost all further training, including the military. Consequently, it makes little difference in one's income whether one has nine, ten, or eleven years of schooling. An extra year at this level, according to the certification theory, has little affect on income. Likewise, it makes little difference whether one has one, two, or three years of college, since without a diploma most opportunities for higher-paying jobs and further education are unavailable. The appropriate analysis for the certification theory is to calculate the average incomes in four educational categories: nongraduates (years seven to eleven), high-school graduates (year twelve), some college (years thirteen to fifteen), and college graduates (years sixteen to eighteen). The results are shown in Table 4-1.

Tables are often thought of as summaries of the data, whereas regression curves are thought of as sophisticated analyses of the data. This distinction is false. They should both be thought of as analyses appropriate to different theories. By themselves, they aren't objective representations of reality. Table 4-1 shows that within the framework of the certification theory, a high school diploma is worth $10,200 a year, a college diploma is worth another $14,300, but some college is worth little more than no college. These numbers shouldn't be taken too seriously.[8] The data set is small and excludes incomes over $100,000. My point is that data don't speak for themselves. They speak only within the framework of a theory.

The major difference between a physical theory, such as Planck's theory of black-body radiation, and a social theory, such as the certification theory, is their scope. Planck's theory not only applies to all luminous objects in the universe, whether a candle, the sun, or the Big Bang itself, but the constant h in the theory plays a critical role in all quantum phenomena. By contrast, the credential theory is limited to human behavior in a particular society in a partic-

Table 4-1

Average Income for Various Categories of Education

Years of Education	Average Income
Less than 12	$14,500
High school	$24,700
Some college	$25,400
Bachelor's degree and more	$39,000

Note: Data are from Fig. 4-1.

ular time. The very terms of the theory, "income" and "education," have no meaning for tribal peoples, or, indeed, for any peoples before the modern era. With such a narrow range of applicability, the theory is difficult to confirm beyond reasonable criticism. It does appear to be, as constructivists would say, just an arbitrary construction that may be useful for thinking about income and education but which doesn't have the potential of being true or false.

Constructivists have taken this dim view of social theory and, by a process of immense overextension, applied it to physical theory as well. I would like to do the reverse. Taking the modest realist position that scientific theory is, or can be, an approximate description of an independent reality (Chapter 2), I would ask: what is required of a social theory for it to be some approximation of reality? The answer, I suggest, is that it must be part of a much larger theoretical framework in which social concepts are seen as particular examples of more universal concepts. Income, for instance, might be seen as a particular resource and education as a particular strategy used in the competitive struggle of members of a particular social animal. Or education might be linked to intelligence and intelligence to physiological conditions in the brain. Although this latter approach is politically controversial at this time, it's the sort of broadening of the conceptual framework that's needed if the social sciences are to be about anything other than their own constructions.

In short, social theory can be just as scientific as physical theory if human beings and human behavior are viewed as part of the natural order. Otherwise, social science is little more than census taking.

IQ Research

The Bell Curve (Herrnstein and Murray, 1994) has brought to the public's attention the field of IQ research that, for a century now, has been measuring the cognitive abilities of large populations. At the heart of this work are tests—usually, but not always, multiple-choice—that can be administered to large numbers of people. Then a mathematical procedure—regression analysis—assesses the degree to which the test scores are correlated with other measurable parameters of life—academic grades, income, and good citizenship. Although the mathematical basis of the analysis is somewhat arcane, statistical software makes it possible for researchers, otherwise ignorant of mathematics, to easily perform regression analyses on personal computers. The software is indifferent to the quality of the data, and can compute the correlation between snowfall and unemployment[9] as readily as between education and income.

To make a science out of a CD-ROM full of data, one must have a theory, or at least the germ of a theory—a hypothesis. Perhaps the most successful, important, and controversial social-science hypothesis has been that concerning standardized tests. For our purposes, it can be stated this way:

> *IQ Hypothesis.* The score on certain specially designed tests measures an essential mental capacity of individuals, called "general intelligence" or *g.* These scores, when scaled so that 100 is the average score of a large population, are called IQs.

Just as "work" in physics means force times distance, and not what one does for a living, so "IQ" means the score on a test, and not how smart one is. The hypothesis is that these scores do measure a real mental ability that is closely described by the common meaning of "intelligence," but what is actually measured—the test scores—shouldn't be confused with what it is that the score claims to measure—a general, and quite stable, mental capacity. Such a claim must be justified by scientific investigation.

If the IQ hypothesis is true, then IQ becomes to the social sciences what atoms and genes are to physics and biology—a fundamental theoretical concept. The consequences of this are considered in Chapter 9. Here I just want to show how the study of IQ broadens the framework of social theory in general, and educational theory in particular.

The Bell Curve investigates the correlation between certain test scores and the likelihood that young white adults will exhibit certain negative social behaviors, such as crime, illegitimacy, and not finishing high school. The simplest way to do this would be to divide the population under study—the database[10]—into (say) five groups according to IQ and to calculate the percent of individuals in each group who (say) didn't finish high school. The result would be a short table showing that nearly all individuals in the highest IQ groups finish high school, whereas a large percent of those in the lowest IQ group don't. The objection with this approach is that the different IQ groups contain different proportions of individuals with high and low socioeconomic status (SES). The lowest IQ group, for instance, contains a much higher percentage of low SES individuals than does the highest IQ group. It could be that SES, and not IQ, determines the likelihood of not finishing school.

One way to sort this out would be to select from the database only those individuals with average SES; that is, to exclude from the study the rich and the poor. The advantage of this approach is not only its simplicity, but its intelligibility. Everyone can understand it. Alternatively, if one has a theory of how intelligence and socioeconomic status affect behavior, one can use all the data. As is discussed in Chapter 2, a strong theory allows many disparate events—oscillating rods of different length, for example—to be considered repeats of the same phenomenon. Since there isn't a theory for how the likelihood of negative behaviors depends on IQ and SES, Herrnstein and Murray model this dependency with a certain mathematical formula that, by adjusting three numbers, can be fitted to the data. This is analogous to fitting a suit to an individual. The tailor can adjust the measurements of the suit to fit an individual human being, but not a fish or a four-armed Martian. Likewise, the form chosen by Herrnstein and Murray can be adjusted to fit a wide range of data, but it isn't guaranteed to be suitable in all conceivable cases. What we have here is a model, not a true theory.[11]

Nevertheless, insofar as the model is a reasonable match to the data, this approach determines a mathematical formula for how the likelihood of not finishing high school depends on both IQ and SES. In this formula, IQ and SES are expressed as standard deviations from the average. An average IQ or SES is zero, and an IQ or SES one standard deviation above average is one. Then, to get at the affect of IQ alone, the formula can be plotted against IQ with the SES variable set at zero (average). This is done in Fig. 4-3. According to this plot, a person with an average IQ has a 5 percent probability of not finishing high school, whereas a person with an IQ two standard deviations below normal has a 60 percent probability of not finishing.

Using the same formula, we can plot the likelihood of not finishing high school against socioeconomic status, with IQ set at zero (average). This also shows an increasing likelihood of not finishing high school with decreasing SES, but not as strong as with IQ. An individual with average IQ and a SES two standard deviations below average has a 20 percent chance of not finishing high school.

Care must be taken not to take any of the numbers generated by the formula too seriously. The formula, after all, uses just three adjustable numbers to fit thousands of individuals. It's like a tailor fitting a suit with one measurement.

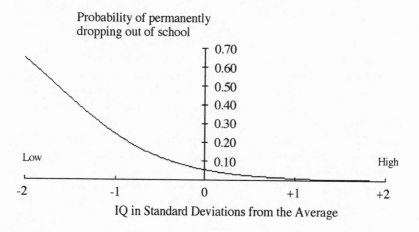

Figure 4-3 Probability of permanently dropping out of high school for whites. (From Herrnstein and Murray, 1994).

The result is a toga. And like a toga, the Herrnstein-Murray formulas give only a crude image of what the underlying body looks like. That is to say, the Herrnstein-Murray analysis gives the rough contours of how IQ and SES affect not finishing high school and a number of other negative behaviors.

The further significance of IQ research on education is discussed in Chapters 9 and 10. For now, it's only being used to show that social science research, in spite of all its inherent difficulties, can objectively investigate an important aspect of human behavior. Since IQ scores are undeniably a property of the brain, and brain development is undeniably under genetic control, IQ research is grounded in basic science. All the work may not be valid—scientists make mistakes—but the work is capable of self-correction and self-improvement. This requires investigators who have strong backgrounds in mathematical analysis and who are disposed to do disinterested research, independent of current academic fads. These habits of mind aren't currently being inculcated in sociology departments, which often pride themselves on their political commitments. They use social constructivism to explain why they aren't doing their share of the heavy scientific lifting.

Ideology is also why educators can advocate addressing the different learning styles of students, while avoiding the whole question of intelligence. Intelligence ranks students, learning styles don't. Ranking is bad, difference is good. Wishes are facts, and ideology is reality. Educational theory, weak to begin with, has completely sunk into the swampy waters of postmodernist nihilism and relativism. And there are few educators who know enough arithmetic to balance a checkbook, let alone understand a multivariant logistic regression analysis.

Dynamical Theories

As important as intelligence may be in understanding individual behavior, it's still a form of categorization, as are the various learning-style classification schemes. Even the chronology and the continuity principles only help us organize our descriptions of events: They tell us the what and when, but not the whys and wherefores. For these we need causal, or "dynamical," theories, that state the mechanisms that govern change. For centuries, the epitome of such a theory has been Newtonian mechanics, which relates the change in the velocity of objects to the physical forces acting on them. Yet the Newtonian program of predicting the motion from the forces works only in rather special circumstances. In most cases, although the Newtonian equations are formally deterministic, they have solutions which are so sensitive to initial conditions as to render precise prediction impossible (Chapter 3).

Outside of the physical sciences, Darwin's principle of natural selection is the best known example of a dynamical principle. The principle of evolution itself—that all species evolve continuously from earlier species—is a corollary of the continuity principle. Natural selection is the postulated mechanism of the observed evolutionary changes. It has been immensely powerful in helping us understand evolution in general, but disappointingly barren in helping us understand the particulars. Basically, evolution is driven by so many strongly interconnected factors that prediction is impossible.

The central question is whether human affairs can be understood in causal terms. That is, can the philosophy and terminology of dynamics be usefully applied to history and sociology? The question is important because public discussion about public policy involves the constant advocacy of this or that action to produce this or that effect. Thus we want to know, in principle, whether such advocacy can ever have a scientific basis.

To the empiricist and skeptic David Hume (1711–1776), a causal connection was no more than the customary expectation—"*not* founded on reasoning"—that if in past experience the cause A always produced the effect B, then a future occurrence of A would produce the effect B. That is, if we have observed many instances of A, say $A_1, A_2, A_3, \ldots$, each of which was followed by an instance of B: $B_1, B_2, B_3, \ldots$, the assertion that A causes B is just a customary way of speaking and reveals no inner workings of nature, no "secret powers." (Hume, 1748/1939)

But no two events are identical; A_1 is never exactly the same as A_2, nor B_1 the same as B_2. This is true not only for events in history, and in human affairs in general, but even for very simple physical systems. We get around the problem in physics by identifying the factors that make up an event. In the case of the rod swinging from a pin, these would be its length, the diameter of the pin on which it is mounted, the diameter of the hole through the rod, and perhaps a few others. We didn't know that all these factors were involved when we started the investigation, but because they played roles of decreasing importance, we were eventually able to sort things out (Chapter 2). Once this was done, the scope of what constituted an instance of A was enlarged by a general formula (law) that correctly predicted a particular B (the time for a particular rod to complete one oscillation). Whether the formula is called a law of nature or a convention depends on its scope. At a certain level of generality, the term "law of nature" may well be justified—by conventional usage.

It could be that human affairs are causal in the Newtonian sense of events following events according to general rules, but that these rules don't lead to repeatable events because the systems they describes are chaotic. In mechanics, even very simple causal systems can quickly become too complicated to predict. For example, the rod that behaves predictably when swinging from a pin exhibits chaos when it is suspended from a wire harness passing through the hole closest to the center (Fig. 4-4). The harness swings on two short pieces of string clamped to a horizontal bar. When the harness is given a strong push, the harness swings back and forth, and the rod sometimes rotates completely around the harness, randomly changing the direction of rotation.[12] The slightest difference in the force of the initial push results in very different patterns of rotations of the rod. The ease with which chaos arises in simple mechanical systems suggests that human affairs, with all the many interacting factors involved, could likewise be both causal and unpredictable.

The recent civil war in Rwanda provides an example of this. When the rebel Tutsi forces defeated the government's Hutu army, it was predictable, perhaps,

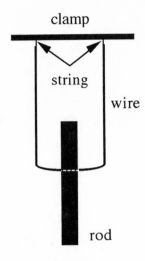

Figure 4-4 A rigid rod suspended in a wire harness will display chaotic motion when the harness is given a strong push.

that millions of Hutus would flee. Relief workers did anticipate this, and established centers in the region the French were protecting, this being the logical place for the Hutus to seek protection. The refugees went, for still unexplained reasons, to Zaire, where nothing was in place for them. Thousands died because relief agencies couldn't predict the exact route fleeing people would take.

The analogy with the rod and harness is strong. In both cases, much of the general course of events is predictable, implying that they are governed by underlying rules and regularities. But some critical events are unpredictable. In the case of the rod and harness, slight differences in the combined motion of the rod and harness can cause the rod to rotate one way or the other around the harness, leading to two completely different histories. In the case of the refugees, differences in roads taken can lead to death or survival for thousands of individuals.

It is thus possible to think of human affairs as being governed by causal rules, yet nevertheless being unpredictable in many instances. The difficulty of developing a dynamical theory of history and sociology is that the theory must be known in order to determine the relevant factors in any situation, and the relevant factors must be known in order to develop the theory. This circularity afflicted physics for thousand of years, yielding only to the genius of Galileo and Newton.

Although the laws of mechanics apply to all objects on earth and in the universe at large, in Newton's time they were applied with the greatest precision to just six objects—Mercury, Venus, Earth, Mars, Jupiter, and Saturn. These objects represented nearly ideal situations: each was a single body moving under the influence of a single force. Once established for these cases, the laws could be applied to more complex situations. The motion of the moon, for instance, is governed by both the force of the Earth and the sun.

In human affairs, there are no "one-force" systems, let alone "force-free" ones. Even the simplest preliterate communities, living in isolation from modern society, consist of a number of interacting subgroups based on age, gender, and clan, and are in alliance and enmity with neighboring communities. Although anthropologists have discovered many fascinating customs of such peoples, they have made no substantive contribution to a theory of social dynamics.

Indeed, until very recently, many social scientists doubted that human affairs could be modeled by the same mathematical methods that are so successful in physics. This is beginning to change, as social psychologists better understand the close parallel between human behavior and the behavior of nonlinear physical systems (Vallacher and Nowak, 1994.). In linear systems, a small cause has a small effect, whereas in nonlinear systems a small cause can have a large, and even catastrophic effect.[13]

Although many students have a hard time with introductory physics, the majority of the equations have the simple form

$$y = ax.$$

This is a linear equation in which a 10 percent change in the variable x results in a 10 percent change in the variable y. For example, the distance d traveled in time t by an automobile traveling with constant speed v is

$$d = vt$$

and the voltage V across a resistor R carrying a current I is

$$V = RI.$$

None of the linear relations of elementary physics is exactly true. In an electrical circuit, the resistance of a wire changes with the current in it, leading to a nonlinear relation between current and voltage. Semiconductor circuits are manifestly nonlinear. Fluid flow becomes nonlinear when turbulence sets in, and even Newton's second law ($F = ma$) breaks down for speeds near the speed of light. At these high speeds, the mass depends on speed.

Nevertheless, much of physics involves systems that can be approximated very well by linear equations. Were this not the case, physics couldn't have developed without the computer, and the computer would never have developed without physics. Now, with the computer, it's possible to model nonlinear systems with great precision and beauty. Most of this work is done by physicists interested in turbulence and crystal formation, but social scientists are beginning to apply the methods to problems in social psychology (Vallacher and Nowak,1994).

My point isn't to discuss this work, some of it quite recent, but to argue that standard methodologies of physics can be applied to human affairs. Even as a metaphor, the nonlinear chaotic systems of physics provide valuable insight into human affairs, which are also highly nonlinear and chaotic. We know from physical systems that this implies some unpredictability of the systems' behavior over time, but it also implies that apparently unpredictable systems can have a simple underlying dynamics. The aim of a science of human affairs is to find this underlying dynamics by showing that it leads to behavior similar to that observed in real people.

At present, one can think of several mutually incompatible principles of human behavior. Some, based on religious teachings, place the main dynamics of human existence in human spirituality, in the human craving for oneness with God, or in human beings' continual disobedience to the will of God. Others put the dynamics of human behavior in the macroeconomic conditions of the world, or in the conflict between different economic interests or classes. In all these cases, the dynamics is seen as resulting from an unnatural departure from some ideal state of perpetual equilibrium. Most utopias are visions of an ideal conflict-free homogeneous society in which everyone loves obeying the dictations of the visionary.

Democratic societies, on the other hand, accept change as the norm, and strive to channel this change in accordance with principles of equity. In large countries, with many diverse groups, classes, and special interests, it's accepted that there will be continual struggle among these elements, and, in fact, this struggle is seen as the best guardian of equity.

This pragmatic view of human affairs has had very wide acceptance in recent years, as country after country turns away from the disasters of planned or dictated societies. This is true not just of the collapse of communism in the former Soviet Union and its Eastern Block satellites, but in the acceptance of free markets in Mexico, India, and many formerly socialist African countries as well. This mass conversion in the political thinking of half the world is perhaps the best evidence the world has ever had that human beings aren't infinitely malleable; they can't be molded into whatever pattern of behavior the social theorist or tyrant dreams up for them. There may be universal principles of behavior, and conflict is probably one of them.

Chapter 5 develops a dynamical theory of human development based on general principles of mammalian and human behavior. These behaviors evolved millions of years ago, under conditions very different from those in which we now live. Do they apply today as they may have in prehistoric times? What is the relation of our moral principles to these basic animalistic principles? Are there objective criteria for distinguishing the relative merits of different societies? Can these criteria be of any guide in developing public policy?

5

Loyalty and Rebellion

Scientists generally avoid grand theorizing about the nature of humankind, leaving such nettlesome discussions to theologians and sociologists. There is great wisdom in this. Edmund O. Wilson was subjected to extreme harassment at Harvard for his brilliant book *Sociobiology* (1975, 1995). Intellectual intolerance didn't begin with the persecution of Galileo, or end with the torrents of outrage heaped upon Darwin and Huxley in the nineteenth century and on Herrnstein and Murray today. Cambridge is still only a few miles south of Salem.

The problem is that any scientific description of human behavior appears to offend many deeply held religious and political convictions. To say, for example, that marital infidelity arises out of the male's sociobiological interest in impregnating many females and the female's interest in attracting good genes, offends religious notions about the sanctity of marriage and romantic notions about love and desire. Yet such a statement isn't an apology for promiscuity or a denial of human affection, but a recognition of the importance of sexual competition in motivating human behavior. Such recognition is in complete accord with common beliefs, from Homer and the Bible to the latest soap operas.

The Trojan War, remember, was fought over Helen's willful elopement with Paris, and the *Iliad* is about episodes of that war that arose from Agamemnon's appropriation of Achilles' mistress. In most stories, disaster follows the violation of a strict sexual code of conduct. Even Romeo and Juliet, who get married before sleeping together, die because they married without parental knowledge.

All societies have rules governing sexual conduct and all societies have trouble enforcing them. The Catholic Church must grant ever more annulments in lieu of recognizing divorce. Is this a breakdown of family values, or an adjustment to a world in which women no longer die in childbirth before they are forty? The purpose of a scientific study of human behavior isn't to undercut morality, or to dictate conduct, but to enlarge the scope of discourse about such matters. It should offend only those who are completely comfortable with their sect's, or cult's, or party's peculiar view of the world.

Principles of Human Behavior

How are the principles of human behavior to be established scientifically? The process, as in the natural sciences, must be iterative. One must assume some principles, work out their consequences, and check the consequences against observations and experiment. If disagreements are found, the principles must be refined or replaced, and the analysis repeated. The formulation of the principles begins with a review of what we know about human nature.

For example, we know human beings are subject to the same physical and biological laws that apply to all living creatures. In addition, our species is a social mammal that organizes itself hierarchically. Individuals struggle within their social group for promotion up the hierarchy, while groups struggle among themselves for territory and access to resources, both human and material. I suspect that this last sentence describes the basic mechanism that drives human affairs. That is, the competitive struggles among individuals and groups is the driving force of history, both individual and national. These struggles influence, and are influenced by, technological, social, and religious developments, but the basic game of life is the struggle within and between hierarchies.

To many people, such a view seems harsh and cynical, but that isn't its intent. Rather it's to find a plausible principle that could form the basis of a dynamical theory of human affairs. Competitive struggle exists in much of human life—politics, business, sports, and education—but so do such apparently noncompetitive behaviors as altruism and voluntarism. The question is whether altruism is an independent characteristic of human behavior, or merely a device that some individuals utilize in their competitive struggle. After all, society honors and supports its holy men as much as it does its warriors; in the United States, some evangelists earn more than generals.

Maternal love is certainly a natural form of altruism and it must be included in any theory of human behavior. It's included here in the statement that "human beings are subject to the same physical and biological laws that apply to all living creatures." All mammalian mothers care for their young, and human caring is a part of this natural order. Paternal caring, on the other hand, probably isn't natural, but is a socially learned behavior. The interesting question then isn't why so many fathers don't care for their children, but why so many more do.

Social Structure

Our species, *Homo sapiens,* evolved some 200,000 years ago from *Homo erectus,* a large-brained, highly developed hominid that had first appeared more than a million years earlier. For all but the last 10,000 years, *H. sapiens* and its ancestors were hunter-gatherers, who moved about in small groups in search of food. Archaeologists have excavated the remains of *H. erectus* campfires from 700,000 years ago, indicating that by then they had discovered the control and use of fire.

These early bands of hominids are believed to have numbered no more than thirty or so individuals, headed in all probability by the oldest male. Life expectancy was low, so the leader seldom was more than thirty-five years old. The women would have been constantly pregnant, dying sooner or later in childbirth.

A long as the economy was based on gathering food from trees and plants, as with modern gorillas, the social structure was simple. Each individual was responsible for his or her own food, and the main prerogative of the leader was mating rights with the females. This necessarily resulted in tension with the younger males. They may have had a few females of their own, but whatever rules governed their rights, we can be sure they were frequently violated: the leader mating with a younger male's female, and vice versa.

Parallel with this tension ran the absolute need of an individual for the emotional and physical protection of the group. Only the group could scare off a leopard or another human band. The solitary outcast would almost certainly die in the dark and hostile jungle which lay just outside the light of the small campfire.

Thus, at the very core of human society, we have a social order based on tension between the need for cooperation among group members and the inevitability of rivalry among the males. This makes life unpleasant for some males, who sooner or later leave the band, taking their females and children with them. As the rivalry between queen bees splits the hive, so male rivalry splits and disperses human bands. This is highly beneficial because it spreads the species geographically and prevents inbreeding.

Unlike bees, however, human groups can grow by mergers and acquisitions. Small, recently splintered groups might fight among themselves for females or merge for protection under an undisputed leader. Females wishing to escape the oppression of an older male might, when opportunity presented itself, elope with a younger one. In this way, the human species spread geographically, while maintaining a healthy mixing of the genes.

And the mixing of genes is, after all, what sex is all about. This is discussed further in Chapter 9, but for now it's important only to understand that there are all-female species that can and do reproduce asexually. Such clonelike species are successful for awhile, but are probably more susceptible to complete extinction when faced with an extreme environmental challenge. A wellmixed gene pool to protect against extinction comes at the expense of a complex and tumultuous mating game.

The dynamics of human behavior is driven by the contrasting needs to be a cooperative member of a communal group and (if male) to compete with members of the group for dominance. This could be called the yin-yang model, with the female yin principle representing community, cooperation, and loyalty, and the male yang principle representing rebellion and rivalry. As in

Chinese medicine, a healthy organization or society is one with the proper balance of yin and yang, of loyalty and rebellion.

Since the rivalry is primordially between father and son, it is masculine. It isn't Oedipal in the strict Freudian sense, however, because the rivalry is for women in general, not for one particular woman. Keep in mind that under the conditions of early hominid existence, an individual would see very few fellow beings, and still fewer nubile females. The father might well control them all—mother, aunts, and sisters. Only much later, with larger human settlements, would there no longer be an absolute scarcity of females. Then rivalry would be for the power and wealth that gives control over the most desirable women.

Females participated in all of this through elopement and marriage. If some weren't willing—even eager—to leave the dominant male, new groups of young males would be unable to acquire a sustainable number of females. In this model, traditional marriage is a grudging accommodation that fathers had to make as human life became more settled and organized. In the West, women won the undisputed right to choose their own husbands only in the twentieth century, as part of a two-hundred-year-long revolution in the status of women in human society. Like the Scientific Revolution that preceded and facilitated it, the Feminist Revolution is a break with hundreds of thousands, if not millions, of years of contrary thought and practice. I know feminists get impatient with the rate of progress—200 years is long compared to a human life span—but in terms of the time scale of human existence, the rate of change has been awesome.

Much of this pattern of group dynamics persists in modern families, where parents try to control their children's sexual activity, and arguments between children and parents over money for clothes can be interpreted as a struggle over the wealth needed to attract the opposite sex. Sons' rebelliousness and daughters' adventurousness are centrifugal tendencies that compete with the centripetal tendencies of communal harmony. These tendencies have different weights in different societies and groups. In black urban society today, the yang of rebelliousness frequently wins, leaving the family without a dominant male. Street gangs act as outcast groups of males struggling to control females and territory. The proximity of rival groups, which are physically much closer than in preurban times, makes it impossible for any group to achieve more than transient dominance.

In most cultures, however, the yin of communal harmony is strong enough to hold males together in the family, however precariously. In Spike Lee's movie *Do the Right Thing,* an Italian father's dominance over his sons, which allows the family to operate a pizza parlor in a black community, is contrasted with the lack of black family cohesiveness and entrepreneurship. The Italian sons aren't happy with their subordinate position, but it's clear that they couldn't make a living outside the pizza parlor. Subordination, however unpalatable it may sound, promotes group prosperity. Actually, most of us are happiest and most productive when we have good leaders. And good leaders need good subordinates.

Modern life offers many opportunities for rebellion and rivalry in business, sports, and science. When an employee leaves a high-tech company to start his own, he is rebelling in the same sense as a tribesman who goes off to start a new settlement. Likewise, scientists who break with current thinking to adopt new theories are rebels. In a major study of over 3,000 scientists who took part in twenty-eight scientific controversies over the past 350 years, Frank Sulloway (1996) has investigated fifty factors, such as age, religion, class, and so on, that might predict whether an individual would be loyal to the old idea or support the new. Of all these factors, the most predictive was birth order. New ideas that were ideologically radical, such as evolution, were accepted much more readily by later borns than by first borns.

Many birth-order studies have suffered from poor methodology, giving the field a tainted reputation. But in a meta-analysis of hundreds of sound studies, Sulloway has found a consistent correlation of personality with birth order (1995). First-borns tend to be more extroverted and less open-minded than later borns. This pattern persists even among the sibling scientists in Sulloway's database. This work strongly supports the idea that the loyalty-rebellion dichotomy is a basic one in human nature, closely linked with rank and hierarchy.

Pure-Case Scenario

But if rivalry and hierarchy are basic to the human social order, they certainly manifest themselves in different ways. Is there a pure-case scenario of human development, perhaps imaginary, of which all real scenarios are modifications? The pure-case scenario would describe, as does Newton's first law of motion, how things would be in the absence of perturbing forces, even though there always are perturbing forces.

My candidate for this idealization begins with a small band of ancient human beings settling in an unpopulated region. Male rivalry within the band is easily relieved by a group of young men and women forming their own group a few miles away. That is, the primordial social order can't maintain groups above a certain size against the centrifugal tendencies engendered by male rivalries, and in this idealized scenario there are no territorial or human barriers to expansion. Thus, as long as there is population growth, groups will continue to split from groups. After several generations, there will be perhaps a dozen groups scattered in a radius of ten miles around the founding group.[1]

For a while these groups share a common language and common customs and maintain friendly relations by exchanging women in marriage. But rivalry and enmity will also develop among the groups, arising from spontaneous fights over women and property. As groups continue to be formed farther and farther from the founding group, a patchwork of shifting alliances and enmities develops that locks individuals into enclaves of allied groups. People have hostile relations with neighbors five miles away, and may never come in contact with people ten miles away. The result is that the separated groups develop their own languages and customs.

Trade goods can travel from group to group, over great distances, without the traders meeting, or even knowing of each other existence. Genes are also exchanged through the marriage of women to neighboring groups. This exchange diminishes with distance, so if the population spreads over a large enough area, racial differences will emerge that increase with distance.

This idealized scenario assumes no contact with other human populations, no limit of land and resources, and no innate tendency for conquest and consolidation. Quarreling and fighting are perpetual—they are the pressure that drives the expansion—but they are self-limiting. The hierarchical structure extends to no more than a few ranked males controlling half a dozen women. If alien beings were to suddenly stumble upon such a world, they would assume that the inhabitants were utterly incapable of organizing a civilization, let alone building spaceships.

New Guinea Highlands

Such an idealized world actually exists in the fertile highland valleys of central New Guinea. When and from where the inhabitants came are unknown, but archeological evidence suggests that people have raised yams and pigs in these valleys for at least 9,000 years, making it one of the oldest agricultural communities in the world. At elevations of 5,000 to 6,000 feet, the climate is mild and springlike throughout the year. Gushing rivers and abundant rainfall are ideal for the gardening that sustains a diverse population of nearly one million people. Most of the tropical diseases that plague the coastal Papuans are absent in the temperate highlands.

Nestled between two mountain ranges, the highlands are completely isolated from the coastal populations. The existence of the populated valleys were unknown to the Papuans and unsuspected by the governing Australians. Australian contact with the highlanders came by accident, when, on May 26, 1930, the prospectors Michael Leahy and Michael Dwyer scaled the Bismarck mountains in search of gold, and found an inhabited, grass-covered valley instead.[2] During the next five years Leahy led a number of expeditions into the highlands, setting the pace and defining the character of the Australian penetration into the region.

> He was the archetypal white adventurer—ruthless and determined, searching for riches in an alien landscape among alien people, driven on by the lure of gold, sustained by an unshakable conviction in the validity of his presence and purpose—a twentieth-century conquistador, arriving just as Europe's great colonising explosion had all but subsided. (Connolly and Anderson, 1987)

Leahy was also a tireless diarist and photographer, whose notes and picture vividly tell the story of the highlanders at the moment of their great encounter. In the 1980s, Bob Connolly and Robin Anderson retraced Leahy's expeditions, and heard the firsthand stories of some of the elders who were present when Leahy first visited their villages (1987). From these accounts, it's clear that the highlanders were eager to marry their women to the men in Leahy's party, both for the bride-price (five pearl shells and a steel axe) and to keep the wealthy strangers among them. Many of the coastal New Guineans in the party did settle in the highlands with highland wives, as did the Australian government agent on Leahy's expeditions. Leahy's brother Dan took two wives, and lived the life of a highlander big man. Michael Leahy married a European woman, and never acknowledged his two sons by native women. Papua New Guinea became an independent country in 1975, and Michael Leahy died there in 1979.

A big man is at the top of the highland's limited hierarchy. Wealth, measured in wives, pigs, and shells, is the only criterion, and this wealth is accumulated through the work of the first wife, who raises enough pigs to buy a second wife. The shells come from the coast through mysterious trade routes that kept the coastal and highland people ignorant of each others' existence. When cash came to the highlands, it was first used to buy shells. Now it buys pots and pans, and school clothes for the children. But pigs still retain their transcendental importance.

The bride-price is another source of wealth for the man who has many female relatives to offer in marriage. These deals also promote personal and clan politics. Highlander women marrying Leahy's expedition members, both Australian and New Guinean, was a continuation of this timeless practice. In principle, a woman could refuse a marriage, but in practice the low status of women made this virtually impossible. The women accepted that their marriage was to advance the interests of their families. So strong was this loyalty that they engaged in prostitution with Leahy's party, a practice unheard of in their culture.

One must appreciate that the highlands is, from many perspectives, as close to Eden as you can get on this planet—perfect climate, fertile soil, abundant water. A garden can be planted in a few hours, a thatched house built in a day. Yet for all that, the highlanders were, and are, as obsessed with wealth and status as any other people on earth. Women are totally subordinate to men, and the men aren't at peace among themselves. Tribal warfare is still a constant preoccupation of highland men, many of whom walk about armed with machetes and bows and arrows. Only the primitiveness of their weapons keeps wholesale bloodshed in check. Languages change every five or ten miles, and friendly territory is very limited. Before the coming of the Australians, most highlanders could never travel more than a few miles from home before reaching enemy territory. One dreads to think what the introduction of guns will do to Eden.

As vast as the highlands is—covering thousands of square miles—the people gridlocked themselves into hundreds of tribal enclaves. Over the millennia, gene flow between widely separated groups ceased, allowing groups to develop distinct racial differences. Western highlanders are taller and more broad shouldered than eastern highlanders (Connolly and Anderson, 1987).

The highlands, free from outside influence for thousands of years, free from environmental stresses, fertile, vast, and protected, is my candidate for the nearest approximation to the pure-case scenario. It shows how people will develop under "force-free" conditions. The best support for this conjecture is the remarkable stability of the lifestyle. The highlanders have been selling their women for shells and pigs, and fighting their neighbors for thousands of years. This is a state of dynamic equilibrium, which in the mathematical theory of dynamical systems is called a "limit cycle." A limit cycle is the dynamical state that a system inevitably reaches, regardless of initial circumstances. That is, regardless of who the people were who first reached the highlands, regardless of their particular culture or beliefs, over thousands of years they would necessarily evolve into a patchwork of many small clans, speaking different languages and fighting one another for land, women, and pigs.

Again, it's the stability of this arrangement that is so remarkable. Although one tribe may annihilate another from time to time, no tribe seems to have ever achieved hegemony over neighboring tribes. There was never much of a class structure, no cities were ever built, and there is no evidence that the people made any technological innovations. Their material culture in 1930 differed very little from what it was after they first discovered how to garden and raise pigs.[3] The smoke from their fires still rose unchanneled to the roof of their thatched huts, passing out through the woven sticks.

Modern notions of progress or human inventiveness have little to do with how human beings have lived for most of their existence on earth. Innovations there must have been, of course, but they are departures from the norm of unending tradition and sameness. Such departures aren't spontaneous eruptions of the human imagination; they are major disruptions in the status quo, forced by changing circumstances or individual genius.

Urbanization

About 10,000 years ago, starting in western Asia, a new form of social organization developed based on extensive hierarchies. This high-hierarchical form has proven to be as stable as the traditional low-hierarchical form, and much more dynamic. It's a second limit cycle, a second state of dynamic equilibrium. It may be that no other organizational form—developed so far—can exist for very long. The breakdown of Somalia shows why human societies are either low-hierarchical subsistence economies—so-called primitive societies—or else complex hierarchical states with rulers who command a military force and a bureaucracy of engineers, governors, and tax collectors. Nothing in between can survive for long because unprotected agriculturalists will sooner or later be attacked by nomadic tribesmen or bandits.

The archaeological record bears this out, since the transition from human societies of hunter-gatherer-herders to settled townspeople living in a walled city occurred rather abruptly about 10,000 years ago. The end of the Ice Age, about 18,000 years ago, started a period of gradual warming which produced excellent hunting conditions in western Europe. Herds of reindeer and bison supplied ample food to the skilled hunters. From this period come the exquisite cave paintings of Altamira in Spain and Lascaux in southwest France. Even as early as 26,000 years ago, there were skilled hunters in southeast France, as evidenced by the recent discovery and dating of a treasure trove of paintings from a cave in Vallon Pont-d'Arc. Bison and other animals are the main subjects of these paintings, often very realistic in their depiction, whereas humans are infrequently drawn, and then only as stick figures. Whether for hunting magic or clan rituals, these paintings indicate a very rich cultural life.

Contrary to common belief, the hunting-gathering economy is not a difficult one, at least for men. The men go out for the hunt, while the women are left to mind the children, gather edible vegetables, cook, weave baskets, and in general take care of things. If the hunt is successful, there is much to eat; if not, the community survives on the food gathered by the women. Apparently, at the time of the cave paintings, there was abundant game and life was good. Subsequent climate changes, however, resulted in a decline in the big herds and a need to hunt smaller, more elusive game.

Further south, in western Asia, some groups started to domesticate animals for food, possibly starting with pigs. In addition, people harvested the wild grains that grow in the highlands that run northward from the Jordan valley to eastern Turkey, eastward to Iraq, and then southward to the Persian Gulf. This led to a settled life in small mud-hut villages of 100 to 150 inhabitants, such as Jarmo in Iraq, which had a mixed economy based on hunting, herding, gathering, as well as deliberate cultivation. Nine thousand years ago the people of Jarmo appear to have enjoyed a varied diet superior to that of the current population (Braidwood, 1960). What we call the agricultural revolution is the displacement of this well-balanced economy with one heavily dependent on grain and manual labor.

This Jarmo village of 100 people was probably a bit more structured than New Guinea highland villages, with a slightly more extended hierarchy. There would have been the subordination of the younger to the older, and of women to men. The oldest men, perhaps a patriarch and his subordinate brothers, would have dominated their wives and children. There might also have been a few unrelated individuals, a captive from a raid or the survivor of a lost hunting party, who attached themselves to the group as servants and slaves. This nascent lower class would have been small, however, and would have shared much the same lifestyle as the others. It wasn't, however, stable.

Great mystery shrouds the transition of human economy from hunting-gathering-herding to settled agriculture. Light garden agriculture, such as that in Jarmo and the New Guinea highlands, may have developed independently in many places. More labor intensive grain-based agriculture arose at a few widely scattered regions of the earth—western Asia (ca. 8,000 B.C.), China (ca. 5,000 B.C.), and meso-America (ca. 3,000 B.C.)—and spread from these over most of the earth. In its wake, we find increasing centralization of power, military organizations, and cities. There is no clear understanding of the order of these events, of their causal connection, or even whether they were the same in different regions.

What is clear is that a prosperous village like Jarmo, consisting of about two dozen mud huts, must have been frequently raided by nomads for its stored wheat and livestock. At some point, instead of hit-and-run raids, a nomadic tribe might take over a whole village, subjugating the surviving inhabitants. The new ruling elite would occupy the huts and confiscate the livestock, keeping the subjugated people as laborers and peasants. This subjugated class had to rely almost entirely on agriculture to survive, because their masters would have controlled the higher-quality food supply.

With all the tribes wandering about western Asia 10,000 years ago, a settled group would have had to vigorously defend itself. The village of Jarmo didn't develop into a city, but the village of Jericho, near the Dead Sea, did. From the time of Jericho's first city walls, some 10,000 years ago, to the city's conquest by the Israelites 7,000 years later, the walls were breached eighteen times (Kenyon, 1954), or once every 400 years. This doesn't mean the walls were useless. On the contrary, for every successful sacking, the walls probably resisted dozens of attempts.

Each time the city was overrun, the invaders became the masters of the surviving population, which was put to work building increasingly elaborate fortifications and houses. Grain, which grew wild in soil irrigated by Jericho's spring, became the main food supply of the captives, the masters taking full control of the cattle and pigs. By digging channels and sowing seeds, the amount of grain could be increased to feed the peasant population. This transformation is certainly one of the most important in human history, since it enabled human beings to live together in larger numbers than ever before.

In this scenario, agriculture arises to feed a captive population, which grows with each successful invasion. These captives, and the free inhabitants of the town, are the instant wealth of the next conquerors. Through the enforced labor of the conquered, the rulers are fed and their dwellings and citadels are built. Agriculture, urbanization, and class divisions arise simultaneously from conquest and subjugation.

Traditional accounts by archaeologists see the transition from hunter-gatherer-herder to urban peasant as the result of an accumulation of techniques that more efficiently exploit the environment (Adams, 1960). But urbanization and agriculture did not improve the life of most people. Soon after they arose, large numbers of people were reduced to peasantry, a slavish way of life compared with that of nomadic hunters and herders. Furthermore, up to that time, human beings and their prehuman ancestors, going back millions of years, had never lived in groups larger than a hundred or so. Thus, I see urbanization as arising from conquest and enslavement, a pattern of behavior which has continued into historical times.

Sparta, at one time a wealthy and cultured city-state, subjugated the people of Messenia in the eighth or seventh century B.C. Thereafter the Spartans became a military caste, devoted exclusively to military discipline and control over the Messenians. At age seven, boys were sent to military school where, under the supervision of older cadets, they learned absolute obedience to the state. Married at age twenty, but confined to soldiers' quarters until thirty, they were encouraged to form homosexual relationships (Bowen, 1972). Throughout history, the aristocracy of a country are the descendants of the last successful invaders: the Dorians in ancient Greece, the ancient Greeks in Italy, the Italians (Romans) in France, the French (Normans) in England, the English and Spanish in America, the Arabs in North Africa, and so on.

In Mexico and Peru, the Spanish conquered established cities with subjugated peasants, in the time-honored fashion. The class divisions in Mexico and South America still reflect who was who. In the Carribean and southeast United States there were no cities to conquer or peasants to enslave, so the Europeans turned to Africa for their peasant class. The transportation of slaves was common in the ancient world, but the scale of the African slave trade was unprecedented in history. What wasn't unprecedented was the creation of class divisions through conquest.

This isn't a pretty story, but neither are the stories in my morning newspaper: racial riots in America, "ethnic cleansing" in Europe, clan warfare in Africa, religious conflicts in Asia. People do not live easily side by side, unless there is some superior power to enforce the peace. This view is controversial, I know. Many people believe that human beings are innately peaceful and good, and that it's government or society or civilization that is violent and evil. But the evidence doesn't support such a romantic view.

I'm not purposely espousing a radical view of civilization, but simply acknowledging some obvious facts. Cities have always contained people from many different clans and tribes, people who would be at war with one another if not for the laws enforced by a central authority. Cities have always had extreme class divisions: the ruling elite, minor officials, soldiers, craftsmen, traders, labors, peasants, and slaves. And in spite of its social and economic inequities, cities have always attracted people of all classes because they offered more opportunities and more excitement than villages.

A major challenge is to understand how a low-hierarchical species could develop a high-hierarchical social system. To some extent, the complex social organization of urban life must be a modular arrangement of simpler social units, just as complex protein molecules are modular arrangements of simpler amino acid molecules, which are themselves specific arrangements of hydrogen, carbon, nitrogen, and oxygen atoms. So perhaps we can think of complex societies as chains of low-hierarchical groups linked together by the same dominance–submission forces that bind the individuals in the group. Just as individual oxygen atoms in a protein are affected by only the few neighboring atoms with which they directly interact, so most individuals are affected by only their immediate family members or work colleagues. That is, on a day-to-day basis, in even the most complex organizations, an individual operates within a low-hierarchical group, with one or two superiors and a few subordinates. These groups are linked together through a hierarchy of group leaders; for example, faculty are organized into departments, departments heads are organized into colleges, and college deans are organized into a university. An individual's primary loyalty is to his face-to-face group, and groups protect their members, even at the expense of the interest of the larger organization. Rivalries and turf battles among departments of any large organization mimic the rivalries and tribal wars of preurban societies.

Nguni of Southern Africa

In the 1810s, cataclysmic events in southern Africa resulted in the almost instantaneous transformation of low-hierarchical tribal peoples into high-hierarchical nations. Watched closely by the encroaching Europeans, we have firsthand accounts of some of the most dramatic events in human history (Ritter, 1955). So rapid were the changes that to a physicist's eye it looks like a system spontaneously moving from one limit cycle (the original low-hierarchical polity) to another (the high-hierarchical polity) because of the growth of an instability in the first cycle.

At the beginning of the nineteenth century, the Nguni of southern Africa were divided into a patchwork of hundreds of tribes and clans. The Zulus were just one of these tribes, occupying a territory ten miles wide. A warrior could walk across Zululand in two hours. The basic social unit of the Nguni society was the kraal, a settlement of beehive-shaped huts surrounding a circular cattle corral.[4] The entire kraal was itself surrounded by a circular stockade fence. The patriarch of the kraal built a separate hut for each of his many wives and for his married sons and their wives, until they could build kraals of their own. Cattle was for the Nguni what pigs are for the New Guinea highlanders: food, money, and sacrificial objects.

Each kraal farmed and grazed its cattle on the surrounding land. The way of life was hardly much different from that of ancient Jarmo, except that agriculture was perhaps more systematic and the Ngunis had iron. For untold centuries this anachronism had little affect on daily life, and then, in an instant, it destabilized the entire system.

By 1800, the kraals were organized into wards under the leadership of one of the kraal's headmen, and the ward leaders reported to a paramount leader, the chieftain or king. This had undoubtedly developed to protect the kraals from tribal raids and to settle disputes among the kraals. The king had absolute power which he exercised through a small army recruited from the many unmarried men in the kraals. These warriors lived in a military kraal that herded and protected the royal cattle, which were, in effect, the national treasury.

Tribal warfare up to this time was usually little more than a jousting contest. The warriors from the opposing tribes stood a prescribed distance apart and threw spears at one another until one side ran away. A few warriors were killed and injured and a few were captured, to be quickly ransomed for cattle. It's unclear how far back in time this ritual went, but suddenly in the 1810s things changed. Perhaps, because of population growth, there was an increasing number of young men who needed cattle and land in order to marry. Or the practice of excessive polygamy by the chiefs and wealthy cattle owners could have created a shortage of women and thus a surplus of unmarried men too poor to pay inflated bride-prices. Whatever the case, the Nguni economy was based on the labor of boys who minded the cattle and women who minded the crops. Young men, with nothing else to do, were recruited into ever larger armies by ever-more powerful kings who needed ever-more cattle to feed and reward their warriors. The result was ever-more wars of ever-more ferocity, which quickly destabilized the social order.

In the early 1810s, Shaka, the exiled illegitimate son of the Zulu king, was a warrior in the Mtetwa tribe. Extraordinary in stature and intelligence, he distinguished himself in fighting, dancing, singing, punstering, and military strategy. He realized that in traditional Nguni fighting—where warriors stood in fixed positions throwing spears at one another— the warriors were actually throwing away their weapons. He discarded his awkward sandals to achieve greater speed and rushed upon his enemies, engaging them in deadly hand-to-hand combat before they could throw their spears. Traveling to the isolated kraal of the iron smelters, he had them make a heavy stabbing blade to attach to a short stabbing handle. A few years later, when he became king of the Zulus, he equipped his whole army with this deadly weapon. Already obedient to a rigid social order, his warriors were subjected to Spartan training and discipline. Brave and self-sacrificing, a Zulu warrior's only hope for marriage and social advancement was through valor in combat. Subsequent battles involved thousands of warriors in hand-to-hand combat that, in a few hours, killed thousands on each side. The winner would slaughter any survivors they found, including women and children, burn the kraals, and take away all the enemy's cattle.

In some cases, the surviving remnants of a defeated tribe became nomads of death, traveling across the land and ravishing everyone and everything in their path. This created more refugees, many of whom fled into the expanding territory of the Zulus for protection. Within a decade, Shaka had made Zulus out of hundreds of disparate tribes, incorporating them into a nation that controlled a thousand times the territory of the original Zulu homeland. It's the rapidity of this transformation from a medium to a high level of political organization that's so remarkable and which supports the conjecture that intermediate levels aren't stable.

And this transformation wasn't an isolated event. One of Shaka's generals, Mzilikazi, turned renegade and fled northward with his tribe. There, through conquest, he created the 500,000-square-mile Matabele empire of central Africa. As with all foreign conquers, Mzilikazi's people became the aristocracy of the empire, with the former leaders occupying the second rank (Ritter, 1955).

This story offers intriguing parallels and contrasts to the stories of the New Guinea highlanders and of urbanization in western Asia and North Africa. Southern Africa differs from the New Guinea highlands in two important respects. First, there is much greater mobility on the open African veld than through the densely forested highlands, especially for an army. Second, cattle is wealth on the hoof. A chieftain who accumulated a large herd had a tempting prize for armed and desperate men. Instead of cities to conquer, the Nguni had kraals filled with cattle to steal. Pigs were as important to the New Guineans as cattle were to the Africans, but pigs are harder to steal in large quantities. Thus, it's reasonable to speculate that the concentration of wealth in sufficient quantity to tempt an army is the major factor in the transformation from a low- to a high-hierarchical system.

Shaka didn't build a city, but he did build a giant kraal, a mile in diameter, to hold his 1,200 concubines. His sexual practices purposely avoided impregnation, since he feared that a son might grow to oppose him. Yet he was so sure of his own invulnerability that he dismissed his guards and went about unarmed. At forty-one, he was assassinated by his two half-brothers, having created a nation with a single innovation—the stabbing blade.[5] The Zulus were conquered by the British in 1879, so they never had a chance to develop an urban civilization independent of Western influence.

Technological Innovation

Prior to urbanization, the rate of major technological innovation may have been one every 10,000 years or so, but with urbanization the rate increased to about one every 1,000 years. From 8,000 B.C. to 3,000 B.C., the urban centers in Egypt and Mesopotamia developed the loom, the potters wheel, metallurgy, wheeled vehicles, ship building, and writing. Important improvements in these crafts were made in the following millennia, but the basic nature of the crafts and the rate of innovation changed only gradually until the age of modern science.

Innovation is rare because human beings are extremely conservative in their habits. At most times and in most situations, survival depends on doing what the community has always done. Shaka's innovations were exceptional and had exceptional consequences. In low-hierarchical societies, human adaptability and innovation were generally needed only on the rare occasions when a group migrated to an environment different from the one in which it had originated. We see this in the United States today, where some immigrants, freed from their own traditions, advance rapidly over those Americans who are still mired in theirs.[6]

Urbanization enabled skilled craftsmen to devote full time to their trade and to pass this trade on to their children, thus creating communities of experts. Though small by modern standards, these communities developed and focused expertise to a far greater extent than was generally possible in preurban societies. This concentration of brain power in cottage industries caused the increase in the rate of innovation that began 10,000 years ago.

Allowing for the vagueness of the concept, the rate of innovation remained steady, or increased slowly, until the Industrial Revolution. That is, preindustrial urbanization must be considered a stable form of organized human life. It lasted for 10,000 years, give or take a few thousand, and was replaced within a few hundred years by industrial urbanization. This new state is characterized by a large jump in the rate of innovation and the utilization of mechanical sources of work on a colossal scale. Whether this new state is as stable as the old is problematic, given its dependency on nonrenewable sources of energy.

Stability here refers only to the continued existence of a system of organization, not to the continued existence of components of the system. Clans were wiped out by plague or warfare in preurban times, and cities, such as Babylon and Carthage, were annihilated in urban times. But wars, sackings, exilings, and enslavings didn't destabilize the urban system of organization—with one mysterious exception. The collapse of the Mayan civilization appears to have been a true destabilization in which a high-hierarchical urbanized system of city states reverted, within a few years, to a low-hierarchical tribal system. By the time Cortes reached the Yucatan in 1519, the great Mayan cities had been abandoned to the jungle for 600 years. It's an example of what didn't happen elsewhere. The rarity of such reversions demonstrates just how stable most urban systems have been.

The Industrial Revolution can be said to have started in 1785, when James Watt installed the first steam engine in a cotton mill. The pace of technological innovation had been increasing in Europe since the late Middle Ages, accelerated by the printing press and the explosive growth in scientific knowledge. By the eighteenth century, the rate of innovation was greater than one every 100 years, the Watt engine having replaced the earlier Newcomen engine in about seventy years. In the nineteenth and twentieth centuries, the rate increased to one every decade, so that the human condition was undergoing a number of major revolutionary changes within a human lifetime.

The nineteenth century took great pride in its accomplishments, and lauded Watt as the "one who was destined to work a mighty change in the condition of the human race by the application of his vast genius to the adaption of steam power to the uses of life" (Lardner, 1840). In step with the inventions of power spinning and weaving, steam power revolutionized textile production. For thousands of years, the production and care of woven materials had consumed any free time a woman might have. Whereas a buffalo hide, once cured and cut, could be worn for years, a wool cloak required spinning wool into yarn, weaving yarn into cloth, cutting the cloth to shape, and sewing the shapes together—mostly women's work. Add to this the washing and drying of the clothes, and you have one of the main occupations of women of all classes for all but 200 of the last 10,000 years. In the *Odyssey,* Princess Nausicaa does the laundry, as she tells her father, the king:

> Father dear, I wonder if you could tell them to get me a big wagon with strong wheels, so that I can take all the fine clothes that I have lying dirty here to the river to wash? . . . There are five sons of yours in the palace, two of them married, while three are merry bachelors who are always asking for clothes straight from the wash to wear to dances. It is I who have to think of all these things. (Homer, 1946)

The serious education of women began only in the nineteenth century, when it was no longer necessary for middle and upper class women to do needle work. This started the women's rights movement, a revolution as profound as the Urban and Industrial Revolutions. Whenever I see women busily shopping among the acres of garments displayed at the mall, I remember that only 200 years ago women spent much of their time making their own, and their men's, clothing.

Certain radical feminists see things quite differently. For them, hierarchy and patriarchy aren't original human conditions. They are, rather, degenerate deviations from an original condition of harmony and equality, when the goddess ruled supreme. Science and industry perpetuate the dominance of men over women and Western culture over non-Western cultures. This anti-Western, anticapitalist, antiobjectivist brand of feminism fails to recognize that science is a girl's best friend. Only science developed the machines that freed women from the drudgery of domestic life and the methodology for finding truth that liberates us all.[7]

Males and Females

In the dynamical model presented here, hierarchy is accepted as a fundamental aspect of human organizations at all levels. It's impossible to have a rankless human society because adults outrank children and older children outrank younger ones. Among adults, there is rivalry for rank or status in the existing hierarchy, which can lead to rebellion and schism. At the same time, the hierarchy maintains group harmony and cooperation by giving each individual an assigned place.

The basic principle of the model is that the family, broadly defined, is the prototypical hierarchy and all other hierarchical organizations, no matter how extensive, are linkages of small units that resemble the prototype. This implies that an understanding of modern life may be found in understanding more primal human behavior. It's interesting to see what such an approach says about the different problems males and females face growing up modern.

For males, age is the primary criterion for ranking in the family, and this natural state of affairs prepares them for the alternating subordinate and superior roles that they will play as they advance through the various hierarchies of school, sports, business, and community life. As they reach adolescence, boys are no longer subordinate to their mothers, and need adult males to hold them in a subordinate rank until they can assume adult roles. The extensive education needed for modern life requires the virtual infantilization of young adult males, a difficult and unnatural process. By fourteen, young men are more inclined to join a war party than to attend an algebra class. The tribalism of the urban streets is a natural attraction to young men who aren't in a contravening male hierarchy at home. Recent efforts to help troubled youths by working through gang leaders is consistent with this model. If the gang is the only social order many young men know, than this is the social order that must be accommodated (Nieves, 1994).

The Liberian civil war was started by Charles Taylor, a Massachusetts-educated Liberian, who recruited an army of uneducated country boys. Under his tutelage, fourteen-year-olds with AK-47s overthrew a government and ravaged the countryside. This was more exciting than school, which wasn't available to these boys anyway. In time, the boys grew up, and as peace talks began in Monrovia in 1995, one early recruit, now a nineteen-year-old veteran, told a reporter: "We are so happy that our father [Taylor] is going to Monrovia. But if he can't bring peace this time, I won't fight again. I'm tired, and I need to go to school" (French, 1995).

The situation for adolescent females is quite different. On the positive side, their continued subordination to their mothers helps them tolerate school better than adolescent men. In spite of claims that young women aren't challenged as much in school as young men, the women are outperforming the men in many instances. At one small, all-white high school in Maine, out of 150 graduates in 1994, the top eighteen were women. On the average, from 1984 to 1994, the proportion of females among the top ten graduates of this high school changed from 50 to 75 percent. The example of this one school doesn't prove much, but the principal assured me that other high schools in Maine are noticing a similar effect. Currently, women are performing as well or better than men in most academic fields.

On the negative side, primal female status comes from youth and nubility, and male status is enhanced by marrying off virgin girls for a good price. Thus, the strict code of virginity protects male interests. Modern life abolishes these interests by allowing women to forego the dubious honor of an early marriage and/or pregnancy for education and work experience. Nevertheless, the code did support unmarried women's interests by protecting them from pregnancy as they worked their way into the main hierarchies of society. The sexual revolution blew this protection away, leaving to each unmarried woman the responsibility for maintaining an active sex life, preventing pregnancy, and competing with men for education and jobs. This is a highly sophisticated agenda, and it's remarkable how well most women have handled it. What isn't surprising is that the same revolution that benefitted the majority of women has marginalized a sizeable minority of women who lack the resources required to cope with these new complexities.

Democracy

The words "subordination" and "rank" have a decidedly antidemocratic tone, and, in fact, many hierarchies are maintained by force and violence. But humankind has discovered, much to its own surprise, that democracy isn't incompatible with effective high-hierarchical organization. Democracy's defeat of the powerful totalitarian systems, which in the 1930s seemed destined to rule the world, gives reason to hope that flexibility and tolerance aren't incompatible with strength and efficiency.

The problem with hierarchical organization is that the guy at the top doesn't have a boss. There was no authority in Nazi Germany to prevent Hitler from making a train wreck of the country, and Iraq and Cuba can't remove leaders with equally poor judgment. The ability of democracies to change leaders constitutionally limits the disasters they can create. A quagmire like Vietnam spells certain political defeat for its plotters.

Democracy does more than blow the whistle on failing presidents and prime ministers. It also leads to more objective thinking on the part of politicians. This may seem an overly generous assessment of democratic politicians, and it is for those representing minor parties that prize ideological purity over governing. But for politicians in office, or seeking office, the need to win broad support in the face of challenges from opponents limits extreme hogwash—at least most of the time. There is accountability, which can sober all but the most egocentric leaders.

Most important, democratic societies permit—even encourage—nongovernmental enterprises of all sorts. Some people are threatened by the enormous power of the large international corporations, but would they feel less threatened if this power were all in the hands of a few politicians? The joint stock company may be the West's greatest gift to the world, because it allows ambitious people to accumulate wealth, not by stealing and enslaving in the traditional way, but by selling stocks to willing investors. In exchange, the seller undertakes to create wealth by putting the money to productive use. The idea that wealth is something that can be created instead of stolen is a product of the Industrial Revolution.

The Unabomber, among many critics of technology and industry, dreams of a return to a low-hierarchical existence in which people are enclaved in small autonomous communes. In a manifesto signed FC, this serial bomber contends that once freed from layers of hierarchy, people will be more empowered and less alienated (FC, 1995). How, after all, do those of us far down in a hierarchy avoid feelings of helplessness and inferiority? The answer, of course, is that all hierarchical life is local. At work, most of us are organized into family-size units with a boss, a number of near-equal subordinates (siblings), technicians, secretaries, and perhaps some part-time or temporary employees. Job satisfaction and one's sense of self-worth come in an inordinate degree from the respect of colleagues and the recognition bestowed by the boss.

Years ago, a newly appointed dean was having trouble with his office staff, who felt slighted by his not acknowledging them when he passed them in the office. An otherwise warm and generous man, he couldn't understand why he was expected to continuously greet people he saw throughout the day. I had just read about transactional analysis in *Games People Play* (Berne, 1964), so I asked him if he acknowledged his wife every time he encountered her at home. He did. "So," I exclaimed with the great authority of someone who has just read a book, "your staff expects the same stroking. It's the way people are."

Improved training for managers can improve office morale and productivity, but people also need to learn to be better subordinates. We can't all be alphas, at least not at the same time. But over time we can play many roles, sometimes dominant and sometimes subordinate. For adults, these roles can change from day to day and hour to hour. The mail carrier may be the director of the community play or chairman of his church's finance committee. I can chew out an errant teaching assistant one minute, and get an earful from my chairman the next. Complex societies offer a much wider choice of roles than is available in small enclaved groups.

It may well be that an important characteristic of a healthy personality is the ability to slip comfortably from one role to another as circumstances require. The CEO must drop his aura of authority when he takes a lesson from the golf pro. During the lesson, the pro is the boss. A healthy society recognizes that all status is circumstantial. No one is superior to another by any absolute criteria, such as birth, intelligence, or position, but by the prevailing circumstances. Even the president of the company has to pick up her feet when the man with a broom comes to sweep under her chair.

Of course, "healthy" is an evaluative word, without objective meaning. It's used here to describe individuals and societies of a certain type. These types might not be considered healthy in a traditional Confucian society in which rank and position are much less fluid than in a democratic society. And Confucian values have persisted for thousands of years, far longer than any democracy.[8]

Verification and Prediction

To be scientific, a model must be not only plausible, but testable as well. The principle that the dynamics of human behavior arises from the struggle between the opposing tendencies of loyalty and rebellion is certainly plausible. We see the struggle at every level of society, from the family to the nation. And the principle has broad explanatory power. It explains the rebellion of the son against the father as well as why it's possible to create large hierarchical enterprises. But these explanations only require that human beings sometimes obey and sometimes rebel, and not that loyalty and rebellion governs much of the dynamics of human behavior. To go from a story to science there must be more than plausibility and explanation. There must be prediction. In particular, the scientific verification of a model requires that the model make a strong prediction, such as the exact numerical outcome of an experiment to high precision, the existence of a hitherto unexpected phenomenon, or the impossibility of some process.

The main conclusion of the loyalty–rebellion model is that all societies and organizations, no matter how complex, are variations of the pure-case scenario of fragmenting and proliferating social groups. The mechanism for this is simple. Rivalry among group members causes groups to fragment and the fragments then compete with one another for territory, status, and resources. This competition acts like an expansive pressure, causing the groups to expand territorially, socially, and economically until all the available resources are utilized. As rebellion and rivalry become isolated within and among ever smaller fragments, conflicts become highly stylized and ritualized. From warring tribesmen to quarreling academics, battles may be highly emotional, but they are usually contained within prescribed limits. At this point, the system is in one of two possible stable states of dynamical equilibrium: low hierarchy or high hierarchy. High-hierarchical systems look much like a confederation of low-hierarchical systems that are held together by some mechanism, military or economic, not present in low-hierarchical systems.

There are two predictions of the model. First, all stable systems are either low or high hierarchies. This means that the transition from a low to a high state—or the reverse—will be sudden. A conqueror can transform tribes into a nation in a few years. Second, these systems are balances between the yin of loyalty and cooperation and the yang of rebellion and rivalry— extreme yin or yang societies are rare or transient. Although we can't perform an experiment to test this, over the millennia there have been many attempts to develop extreme societies. These natural experiments show the limits of the yin–yang balance. In particular, the socialist experiments of this century strongly suggest that the rebellious side of human nature can't be suppressed without such tyrannical force as to undercut the viability of the whole society. Perhaps the most successful collective, or yin, society was ancient Sparta, a barracks state so bizarre that we remember it to this day. The whole purpose of the regimentation was to exercise control over an enslaved population, a state of affairs that would have been repeated had Germany won World War II. The kibbutzes of Israel also come to mind as a successful example of collective living. But they are voluntary associations of a tiny fraction of the Israeli population. Like monastic life, kibbutz life is widely admired, but seldom followed.

There is thus some circumstantial evidence to support the prediction that most societies operate in the middle of the loyalty–rebellion scale. Within this limited range, societies differ in their level of tolerance for dissident behavior, from traditional societies with low tolerance to modern societies with high tolerance. In this view, the individualism and autonomy prized in the West are forms of rebellion, whereas Asian societies stress family loyalty and subordination according to rank. In Taiwan, it's the deviant gangster (*liumang*) subculture that values community loyalty over family loyalty and individual autonomy over subordination (Shaw, 1991). Modern Asian and Western societies may well respect the loyalty and rebellion limits that are possible for modern industrial states. Any more obedience in Asia, and the society loses the minimal flexibility and creativity required for adaptability. Any more deviation in the West, and the society loses the minimal social cohesion to survive.

This model views societies and social institutions as complex dynamical systems. Although the detailed behavior of such systems is unpredictable, they evolve toward limit cycles, or states of dynamical equilibrium, in a predictable way. For example, imagine dropping some food coloring into a glass of water. The details of how the color disperses throughout the water is unpredictable, yet it's absolutely certain that it will disperse. Furthermore, after a long enough time, the color will be uniformly distributed throughout the water. The end result is certain, regardless of the details of the dispersal process. This final state is called a state of dynamical equilibrium because the food-coloring molecules continue to move randomly among the randomly moving water molecules. The state of uniform dispersal, which is visually rather ordered, is the most disordered situation from the molecular viewpoint. The system has evolved to the state of maximum disorder, as stipulated by the second law of thermodynamics.

Even more analogous to the social model, imagine simultaneously pouring a glass of water and a glass of oil into a bottle. Again there will be a complex and unpredictable mixing of the liquids, but after a time all the oil will float to the top. At equilibrium, there will a layer of oil floating on a layer of water, because molecules of the same kind are more strongly attracted to one another than are molecules of the opposite kind. In this case, transient mixing can be restored by shaking the bottle.

A beautiful example of this mixing and separating process can be observed by shaking a nearly full bottle of baby shampoo. The shaking mixes hundreds of air bubbles in the liquid, which rise to the top after the shaking stops. There are bubbles of all sizes, and they move at speeds proportional to their size, so that the separating process looks like a dance. Again, much of what will happen can be predicted, though not the exact size and position of each bubble.

Thus in dynamical systems one can distinguish transient processes—the dance of the bubbles—from the limit cycles, the states of dynamical equilibrium. The loyalty–rebellion model predicts that the equilibrium state of low-hierarchical societies is a gridlocked mosaic of mutually antagonistic clans and tribes. This was the state of the New Guinea Highlands in 1930, and the state toward which Somalia may be headed. The equilibrium state for high-hierarchical societies is a gridlocked mosaic of classes, guilds, and ruling organizations such as the government, the military, and the priesthood. This state has been reached many times in history, from ancient Egypt to modern Britain, which has been described as "a cozy backwater, a back-slapping, amateurish 18th century oligarchy, its boardrooms stuffed with has-been politicians, Foreign Office retreads and Establishment cronies" (Denman, 1995).

Coziness is just another term for the ritualization of conflict that occurs during long periods of stable equilibrium. In this state, attention is diverted away from real issues and directed toward maintaining rank and position according to rules that value style over content. The coziness of the American educational establishment is evident in its endorsement of every feel-good doctrine and mandate, and its refusal to make hard decisions among the many competing demands that are placed on it. Fragmented education faculty set course requirements for future teachers as chieftains divide a conquered territory. The system, from the National Science Foundation in Washington to the best school districts in the country, is incapable of developing a simple eighth-grade science curriculum.

Disequilibrium is usually caused by war and migration in the case of nations, and by merger and takeover in the case of organizations. These disasters shake up the system, disrupting the rituals and empowering the innovative and the resourceful. It's probably during these periods of turmoil and disequilib-

rium that new ideas and inventions arise. Since the Scientific and Industrial
Revolutions, innovation has been the cause, as well as the consequence, of disequilibrium. For thousands of years scribes formed an honored guild that controlled the production of books. This monopoly was overthrown by the print-
ers, whose guild of typesetters had considerable power until it was destroyed by
the computer. In the United States, the maintenance of disequilibrium in the
private sector is a matter of public policy, supported by antitrust laws. The 1984
breakup of AT&T is one of the boldest peaceful bottle-shakes in history. It was
motivated by the conviction that smaller, less regulated companies are, in the
long run, inherently better able to adapt and innovate than is a highly regulated
monopoly.

The failure of recent efforts to reform the formidable Japanese bureaucracy
foreshadows a time when Japan might become too inflexible to meet emerging challenges. Walter Mondale, the U.S. Ambassador to Japan, has commented
on "the real obduracy to change here." Japanese officials are afraid that any
movement away from obedience would lead to crime and social breakdown.
They prefer the corruption of business payoffs to politicians—Japanese and
otherwise—to the rebellion of street crime. Or as one official put it, "deregu-
lation would create great confusion" (Sterngold, 1995).

From the point of view of this model, many of the turmoils of modern
life—business restructuring, mergers, migrations, international competition,
elections—are mechanisms that keep the social order in a state of mild disequilibrium, preventing gridlock and stagnation in technology and manufacturing. Public education, as a state-run monopoly, has been sheltered from external pressures and allowed to reach equilibrium. In this condition, it's incapable
of changing itself. Meaningful change requires a shake-up sufficiently violent
to create major disequilibrium. Massachusetts has enacted some drastic measures designed to do this, but it remains to be seen whether they are drastic
enough (Chapter 10).

Interestingly, science itself, thought of as a system of practitioners, both cor-
roborates and contradicts this model. In corroboration, we find many instances
of rebels, such as Sigmund Freud and the parapsychologist J. B. Rhine, breaking from the mainstream to form their own movements. Indeed, the whole social constructivist movement exemplifies the fragmentation and isolation that
is a central prediction of the model. Sociologists and their kin break away from
any allegiance to science, and then the break-aways further split into a babble
of noncommunicating subgroups (Chapter 4).

In contradiction, there has been a remarkable unity of viewpoint within
mainstream science, even as it has grown to encompass millions of practitioners around the world. All of its many subspecialities build upon the same body
of established knowledge, use much the same instrumentation, and accept the
same rules of evidence. This to me speaks of the underlying reality of its subject matter. Of course, particular scientific institutions are as subject to internal
fragmentation and isolation as any other social unit, and whole branches of science, such as sociology, can rebel. The total fragmentation of science itself is
possible, should society as a whole lose its faith in objectivity and reason.

Conclusion

At one level, this chapter is an exercise to show that a dynamical model, governed by a simple mechanism, can have enough complexity to predict some general features of human behavior. The mechanism is selected from among the many factors that influence human affairs for its simplicity and plausibility. Predictions are made on the basis of the model and compared with experience.

At another level, this chapter argues for the truth of the loyalty–rebellion model. This model is pessimistic in that it predicts that societies and social institutions naturally evolve toward a gridlocked state of fragmented and isolated social units, unless the process is disrupted by external, and often catastrophic, forces. This is contrary to our ideal of a coherent and integrated society. The second law of thermodynamics is pessimistic in the same way, since it states the impossibility of building a perpetual motion machine. Yet because of the second law we know how to build engines with nearly the maximum possible efficiency and avoid wasting time pursuing impossible goals. I'm an optimist in believing that knowledge of the true dynamics of human behavior can help us build the best possible society and prevent us from wasting time pursuing impossible goals.

Most educational philosophy in the United States is based on a very different social theory—one that views hierarchy as an aberrant form of human organization. What is the origin of this view? Is American education locked into a set of unrealistic ideals and expectations? How do these ideals affect educational policy? Can a more realistic assessment of what is possible lead to more effective educational policies? Who's in charge here, anyway? Chapter 6 reviews the history of education from ancient times to the present in search of an understanding of America's peculiar "unsystem" of education.

6

A Brief History of Education

For more than three million years, we and our hominid ancestors have been taught specific skills and behaviors. Until very recently, most of this knowledge was acquired more or less naturally, as a consequence of a child's normal activities within a small community. Some was acquired by imitation: a young boy playing with a fire stick, twirling it in a clump of dry moss as he has seen his father doing. Some was acquired by informal tutoring: a mother pointing out to her daughter the subtle differences in the shape of the leaves of the edible and inedible varieties of a plant.

Formal education as we know it today takes place apart from everyday life and is primarily about and through language. There are several problems with this overvaluation of a single mode of expression. First is the illusion of the omnipotence of language, which is the belief that language can express everything. But, as we'll see, language is essentially baby talk. It can't express complex ideas or emotions; these require mathematics, art, and music. By overvaluing language, other modes of cognition are undervalued. Students are expected to study literature in school and to get their music in the streets. The ultimate absurdity of this illusion is the claim by some that the analysis of texts—literary criticism in its broadest sense—is the master science, since all human knowledge is thought to be expressed in language (Carroll, 1996).

Second is the illusion of the magic of language, which is the belief that words have power over matter. This manifests itself when prayers are published in a newspaper in the belief that they will produce material benefits, or when educational standards are published in the belief that they will improve students' knowledge of science.

Third is the illusion of the realism of language, which is the confusion of words with the things they represent, or the belief that words have meaning apart from what they represent. Thus a science textbook introduces the concept of energy by asking the students for examples of things that have energy (BSCS, 1994). This makes no more sense then asking students for examples of things that have entropy. It's no more possible to extract the meaning of energy from the word "energy" than it is to extract the meaning of entropy from the word "entropy." The extreme realism of language found in the BSCS middle-school science textbooks results in their treating dissimilar concepts as similar simply because they have the same name. For instance, technological evolution and biological evolution are treated as similar examples of "evolution," even though their underlying mechanisms are fundamentally different.

That language often muddles thinking more than it clarifies it certainly argues for the need for good language education. But overvaluation of language, which is what teachers are particularly skilled at, can result in undervaluation of other cognitive abilities. A review of the history of language and formal education may help to give us some perspective on why things are as they are.

Language and Thought

We don't know when or how language originated. All of the some 5,000 existing human languages have fully developed syntactical structures; none is primitive in the sense of being in an earlier state of development than another, yet languages are not equal in their degree of complexity. Languages spoken by small intimate populations tend to be grammatically and phonetically more complex than languages, such as pidgins, that are spoken by large multiethnic populations (Hymes, 1974).

Little can be learned about the origin of language from studies of existing languages. At best, linguists have been able to trace families of language, such an Indo-European, back to a single hypothetical language some 5,000 to 10,000 years ago. All the language families may have originated from a single language at some still earlier time, before the species dispersed across the face of the earth (Haugen, 1974). Since human beings reached Australia at least 35,000 years ago, language is surely much older than this. And since language changes radically in the course of a few generations, all trace of its original state was obliterated from adult language tens of thousands of years ago.

But perhaps not from children's language. Language develops in children in stages, and many other developmental processes, both biological and cultural, tend to recapitulate their historical evolution. For instance, the human embryo develops gills before lungs, and science students learn the concepts and principles of Newtonian mechanics before those of quantum mechanics and relativity. Since the 1960s, the painstaking recording of children's utterances have enabled psycholinguists to piece together a universal developmental sequence of early language acquisition in children.

My late cousin, the psycholinguist Richard F. Cromer, was a pioneer in this field. He studied under Roger Brown at Harvard, and continued his work on the acquisition of language in normal and handicapped children at the Medical Research Council in London. Work in this field was motivated by the quest for universalities in language acquisition, and as Brown once wrote, it was this quest that "sustained the researcher when he [got] a bit tired of writing down

Luo, Samoan, or Finnish equivalents of 'That doggie' and 'No more milk'"
(1974).

The main conclusion of this work is that the earliest uses of language develop
spontaneously from a child's innate language-making ability, and are not merely
imitations of adult speech. For example, my nephew Marty's first word was
"ban" which referred to his toy train. It was rapidly generalized to refer to real
trains, and then to mechanical equipment in general. This clearly was not something he learned from his parents, but something he created. He taught the
adults the word, they didn't teach it to him.

This suggests that language, at its very beginning, was an infant's invention
and its original function, therefore, must have been to advance an infant's agenda.
That agenda, of course, is to obtain care and sustenance from the mother in
competition with older and younger siblings. In early times, the average woman probably bore eight to ten children, only two of whom lived to reproduce.
Many factors determined who would survive, most importantly resistance to
disease. But the ability to extract an extra measure of care—to be the most lov-
able—played a role as well. Although this violates the modern Western moral
code in which all children must receive all the care they require, it violates no
law of nature. In a world of limited resources, an infant who can extract more
care increases his chance of survival. And the extra bonding that arises when an
infant teaches his mother his first word may give that child the extra edge he
needs to survive.

Over many generations, this process would gradually increase the percentage of linguistically competent individuals. When there are several one-word
infants in a group, they can start to learn from one another and to help one
another remember their words as they get older. As with children and computers today, children hundreds of thousands of years ago may have been speaking a primitive language that none of the adults yet knew. These children would
use their language with younger children, molding their natural language-cre-
ating capacity to the emerging group language. Even today, parents sometimes
must ask their six-year-old to interpret their three-year-old's speech.

This is in keeping with the revolutionary conception of Noam Chomsky
(1986) that human beings have an innate and autonomous language ability that
functions independently of other cognitive abilities. He reached this conclusion
by analysis of adult language, arguing that it is much too complex to be learned
the way reading and writing are learned. Studies of language acquisition in
children show that by a certain age all children correctly use "un" with verbs
like "tie," "cover," "wrap"—"untie," "uncover," "unwrap"—and refrain from
using it with verbs like "tear," "cut," or "wash." Likewise, they distinguish count
words, like "bottle," "box," and "bean," that have plurals and take the articles "a"
or "the" in the singular, from mass nouns, like "water," "sand," and "rice," that
aren't normally used in the plural and take the quantifiers "any" and "some"
(Richard F. Cromer, 1991). The very subtlety of these distinctions, which native
speakers unconsciously understand, leads to the conclusion that human beings
have an innate ability to perceive linguistic categories. In this view, language
isn't learned, it grows.

For nonlinguists, the most convincing evidence of this remarkable thesis comes from studies of severely retarded children who have normal, and even extraordinary, language ability. Cromer reported the case of D. H., a spina-bifida child with arrested hydrocephalus. Though severely retarded by all standardized nonlanguage tests, and by late teenage still unable to read, write, or handle money, D. H. could speak with great fluency and complexity: "Mum didn't mind me moving about, but Dad objected to it because he knew it was bothering me and it was bothering my school work." There are similar cases of other retarded, but linguistically competent, spina-bifida children (Richard F. Cromer, 1991).

Another type of retardation, called Williams syndrome, points to the same separation of linguistic and cognitive functions. Williams children have characteristic elfin-like faces and are extraordinarily affable, verbal, and musical. Yet their problem-solving abilities are comparable to those of Down syndrome children: they can't tie their shoe laces or make change for a quarter. The trait has been traced to a set of missing genes on chromosome 7, which includes the gene that codes for elastin, the elastic protein found in all connective tissue. On autopsy, the brains of Williams syndrome people show numerous abnormalities, except in the frontal lobes, medial temporal lobes, and the neocerebellum, regions of the brain associated with speech (Blakeslee, 1994; Boucher, 1994; Wang and Bellugi, 1994; Morris, 1993).

If we accept that language is independent of cognition—a tough pill to swallow, I admit—then it follows that language didn't evolve from cognition. Furthermore, that Williams syndrome spares both sociability and speech, supports the hypothesis that speech evolved to improve loveableness, not communication. Its original purpose wasn't to aid in hunting or to teach a child how to make an arrowhead; the original purpose of speech was to talk. That is, talking, in and of itself, is what language is for. Other purposes, such as conveying information, were later adaptations.

Even today, most activities and informal learning take place without the use of language. Although much talking may accompany an activity, little is required for the activity itself. Probably much, if not all, of *H. erectus's* culture was transmitted without language. Most manual skills today are learned by imitation and practice, with verbal instruction playing a very small part. Indeed, language is hardly up to the task of describing even moderately complex mechanical operations, as anyone knows who has tried to decipher the instructions for assembling a barbeque grill or furniture in a box. The instructions for assembling my computer workstation were mostly pictures, with very little text. Although this was quite intimidating at first, I soon picked up the pictorial conventions, and found it easier to go from pictures to objects than from words to objects.

Language is so much a part of education today, that we often fail to appreciate how much important education takes place without it. Carpenters and plumbers, athletes and musicians, learn their skills from the guiding hands of mentors and masters. Words may be used to correct and encourage, but they can't convey the full complexity of the motor and visualization routines involved in such activities. Yet language skills—reading, writing, and speaking—define, by themselves, what it means to be educated today.[2]

Reading and Writing

Formal education doesn't take place in context or in response to the curiosity of the learner, but at a place and time determined by the teacher. Most societies, even preliterate ones, have some formal education, usually as part of the religious and ceremonial training of initiates into secret clubs. In the highlands of Papua New Guinea, young Huli men go off to live in the woods for two years while their hair grows into a frame. At the end of this period, the frame with its hair is cut from the young man's head, and becomes a wig that he will wear the rest of his life. While in seclusion, the wig teacher instructs the young men on the secret lore of their culture, especially the secrets of growing hair.

Education, as we know it today, is intimately associated with reading and writing. The oldest known writings are five-thousand-year-old Sumerian commercial documents inscribed on clay tablets. The system was already highly developed by then, with over 1,500 different pictographs and ideographs. The city-states of Mesopotamia must have had organized scribal schools centuries earlier to develop and teach this complex notational system. Student exercises on clay tablets from 2500 B.C. have been found and even possible classrooms—rooms with rows of stone benches—have been excavated from a 2000 B.C. site (Bowen, 1972).[1]

Writing developed independently in Mesopotamia, China, and Mesoamerica. In China, the oldest known writings are inscribed oracle bones and shells (ca. 1500 B.C.) and in Mesoamerica, the oldest writings are Mayan inscriptions on monuments containing elaborate pictures of kings performing religious rituals (ca. 300 B.C.). Although nothing certain can be said about the original purpose of writing in each civilization, the evidence we do have indicates that it was different in each case. In Mesopotamia, writing most likely originated for commercial purposes, whereas magic and fortune-telling may have been the initiators in China. For the Maya of Mesoamerica, "writing was a sacred proposition that had the capacity to capture the order of the cosmos, to inform history, to give form to ritual, and to transform the profane material of everyday life into the supernatural" (Schele and Freidel, 1990). After the collapse of Mayan civilization in A.D. 900, monumental writing ceased, though literacy continued among the shamans until the Spanish conquest.[2]

The precursor to writing in Mesopotamia has been traced to clay tokens that were in wide use in the area from 8000 B.C. to 3000 B.C. Of a variety of shapes—spheres, disks, cones, tetrahedrons, and so on—they were probably part of an active commercial system, perhaps to record sales of goods and livestock. When enclosed in a baked-clay sphere, or "bulla," they served as a bill of lading, allowing goods to be transported by third parties without fear of theft. Entirely enclosed in a bulla, tokens representing the number and kind of goods being transported were carried by caravan from seller to buyer. Upon breaking open the bulla, the buyer could check that he received what had been sent. To keep track of the goods without having to crack open the bulla, each token was pressed into the outside of the bulla before being encased in it. Many of the earliest pictographs used in Sumerian writing can be matched to the impressions these tokens make when pressed into clay. For example, a disk inscribed with two lines crossing at right angles, when pressed into clay, is a circle with crossed lines, an early pictograph for sheep (Schmandt-Besserat, 1978)

This marvelous analysis by Pierre Amiet and Denise Schmandt-Besserat would seem to clinch the assertion that writing in Mesopotamia originated for commercial purposes. This is very significant, especially if this wasn't the case in China and Mesoamerica, because it indicates that from the beginning of civilization in these areas, the status of commerce may have been quite different. The utilitarian foundation of writing in western Asia may explain why the alphabet developed there.

The earliest scribes in Mesopotamia were little more than inventory clerks, with a prestige comparable to that of a skilled craftsman. But as the system of writing developed, and especially as the need for formal schooling grew, the craft garnered the prestige of a learned profession. By 2000 B.C., Mesopotamian scribes had to read and write both Akkadian—the language of daily life—and Sumerian—the ancient language of literature and religion. They also learned to keep business accounts and to take dictation so fast that, as a four-thousand-year-old copy exercise puts it, his "hand moves in accordance with the mouth" (Bowen, 1972).

The education of a scribe began in a House of Tablets, which prepared students to work in lower-level commercial jobs. The more gifted students continued their education in a House of Wisdom, which prepared them to fill the higher administrative offices of the Mesopotamian city-states. Around 1700 B.C., the Babylonian king Hammurabi succeeded in gaining control over many of the independent cities of Mesopotamia. This Babylonian period developed a rich literature and legal system, some of which has been recovered from libraries burned and buried thousands of years ago.

The Mesopotamian writing system was based on the sound of the spoken word. The unit of sound was the syllable, both open (consonant-vowel, such as "la") and closed (consonant–vowel–consonant, such as "lat"), so that as many as 150 symbols were required. True alphabetic writing, in which each symbol represents a consonant, rather than a syllable, seems to be the unique invention of an unknown people of the Levant or western Asia around 1000 B.C. From this, two major families of alphabets evolved: the Aramaic-Hebrew-Arabic family and the Phoenician-Greek-Latin family.

Greek Education

The Greeks adopted the Phoenician alphabet, adding symbols for the vowels. Prior to this, wealthy families had hired tutors to teach their children to sing and play the lyre, to recite the works of Homer and Hesiod by heart ("rote learning"), to wrestle, and run, and throw the javelin, and to mind their manners. There were probably private schools as well, where for a reasonable fee the sons of the merchant class were taught in larger groups. That is, a market for formal schooling existed prior to literacy, driven by the desire of parents to have their children acquire a high level of cultural attainment. Eloquence of speech was particularly important, since civil life, including legal disputes, depended on oral presentations. The Greek school boy reciting his lesson to his seated master is a common theme on Greek vases.

The full alphabet simplified the teaching of reading and writing, giving the schoolmasters a very important subject to add to their curriculum and further increasing the market for their services. The demand probably grew very fast, not because the average citizen had much use for reading or writing at the time, but because of the need to keep up with the Critos. There was nothing like it in education until 1981, when the personal computer gave educators a new "literacy" to teach.

I was involved with the educational use of computers myself at the time, and so I had a privileged view of this singular historical phenomenon. No sooner had my colleagues and I begun exploring the potential use of the computer for teaching science than colleges began offering master's degrees in computer education. Although no one had any knowledge or experience using computers to teach anything, instant experts were trained, hired, and funded to bring computers into the public schools. It was an educator-driven (and manufacturer-driven) enterprise that parents quickly bought into for fear their children might be left behind. Mass literacy in Greece in the fifth century may have spread with comparable frenzy. Unfortunately, no manuscripts have survived from the critical period between 1200 B.C. and 300 B.C. when alphabetic writing was developed. By the third century B.C., spelling had become standardized in the different Greek dialects and the systematic analysis of writing and language—philology and grammar—were well established. It probably won't take quite as long for computer education to reach a comparable level of standardization.

The word "pedagogue" (*paidagogos,* literally boy's escort) originally referred to the slave who escorted a boy to and from school. The only references we have to these schools are a few accounts of ancient accidents, such as the collapse of the roof of a school on the island of Chios in 496 B.C. that killed all but one of the 120 children in it (Herodotus, 1954; Bowen, 1972). There were undoubtedly many such schools in Athens and in the other Greek cities of Greece and Asia Minor.

In Athens, in the fifth century, there also arose a market for higher education. To meet this demand, teachers from throughout the Greek world came to Athens, where they established schools based on a variety of methods and doctrines. There was no fixed subject matter. Some schools focused on practical matters, such as successfully arguing a case in court, whereas others developed philosophical discourse and analytical thinking. But common to all was the spirit of creativity and open inquiry. Because of the strong tradition of young men exercising in public gymnasia, schools were often established near them, and in time a secondary school came to be called a gymnasium. Plato's school was near a grove dedicated to the shrine of Acadmus and so came to be called the Academy; Aristotle's school was called the Lyceum after the name of a nearby gymnasium.

By the third century B.C., scholars in both China and the Hellenistic world were deeply involved with the study of their respective classical traditions. Schools founded on the classical works compiled by Confucius in the sixth century B.C. became, in the third century B.C., the official schools for educating government officials. Similarly, higher education in the Greek world developed for a time along lines laid down by Athenian philosophers of the fifth and fourth centuries. But in time, education in the East degenerated into excessive literary analyses—the illusion of the omnipotence of language—and by the fourth century A.D. education in the West had become formalized "masses of information that had to be learned even if their relevance was not understood or their values given application" (Bowen, 1972). From this deplorable state, education in the West declined to near extinction, with a limited literacy surviving among a few Christian monks of the sixth to tenth centuries, as it had among a few Mayan shamans of the tenth to sixteenth centuries.

Medieval Education

By the eighth century, literacy in Europe had declined to such an extent that even the aristocracy were no longer taught to write. Letters, legal documents, and accounts were all written by monks, who were the sole civilizing force in a barbaric age. Recognizing the dire state of knowledge in his time, Charlemagne (742–814) recruited the most learned monks to his court, providing them with the means to establish schools for boys from all classes. "In this way," wrote his contemporary biographer Einhard, "Charlemagne was able to offer to the cultureless, and, I might say, almost completely unenlightened territory of the realm which God had entrusted him, a new enthusiasm for all human knowledge" (Bown, 1975).

The revival of education in Europe was slow and unsteady, hindered by a chronic lack of qualified teachers. Monasteries often restricted education to their own initiates, though by the tenth century schools attached to cathedrals became genuine centers of scholarship. There, original thinkers like Peter Abelard (ca. 1079–1142), could, at great risk to themselves, raise troubling questions about the nature of God and the meaning of words. Indeed, language was central to the developing scholastic tradition, based as it was on the few works of Aristotle known at the time—principally *Categories* and *On Interpretation*—which treated the method of seeking truth by discussion and correct reasoning. Aristotle's scientific works became known only much later, while belief that language is the sole mechanism of cognition continues to this day. Postmodern academics, led by Jacques Derrida and Michel Foucault, have even managed to turn the argument inside-out: since language is synonymous with thinking, and language is incapable of fairly representing reality, objective knowledge in general, and scientific knowledge in particular, is impossible (Gross and Levitt, 1994).

That the black hole of textual analysis didn't engulf all of learning is one of the miracles of Western civilization. This is no doubt because medieval trade guilds developed a separate educational system to teach their children the practical arts. Central to this was the apprentice system, which indentured a boy at thirteen to a master for seven years of work and training. Prior to his apprenticeship, the boy received some elementary education in schools run by the guilds. Afterward, the young man became a journeyman, advancing to the rank of a master upon completion of a major work, or masterpiece. This system maintained the competency of the guild's craftsmen, while controlling access to the trade.

In the twelfth century, teachers at the schools associated with cathedrals began to organize themselves into guilds for the same reasons. At the time, *"universitas"* meant the guild itself, and *"facultas"* referred to the subject divisions of the guild: law, medicine, theology, and arts. Later, "faculty" became the term for the guild of masters, and "university" for the corporate institution of masters and students. It is remarkable that the modern university, from the training of doctoral students to the organization of the faculty into departmental enclaves, is still modeled on the medieval guild system.

But the craft guilds gave more to education than its organizational form. By the sixteenth century, practical treatises on technology became available through the medium of printing—itself a medieval technology. This bridged the age-old gulf between the theoretical knowledge of the academy and the empirical knowledge of the artisan, making the connection that stimulated the development of modern science. Of course, the vast majority of academics ignored practical knowledge, trained as they were in rhetorical argumentation. But a few original thinkers, like Francis Bacon (1561-1626) and Galileo Galilei (1564-1642), understood that all knowledge was unified, so that theory and practice were allies in the same struggle to understand nature. Galileo built his telescopes from the knowledge of lenses and lens-making which the optical trade had been developing since the invention of spectacles in the late thirteenth century (Charleston and Angus-Butterworth, 1957). With his discovery of the moons of Jupiter and of the craters on the moon, Galileo showed that the natural world contained wonders that could not be seen with the unaided senses. This downgraded the role of the scholars of ancient texts in understanding nature, and upgraded the role of the makers of instruments.

More generally, the age-old dichotomy between Platonic rationalism and Aristotelian empiricism was resolved when Newton showed the immense intellectual power that resulted from combining reason and observation, just as Bacon had predicted. As obvious as this may seem to us now, it had been generally denied up until then because it undercut the basic religious principle that knowledge of God comes from intuition. Religion must deny the importance of empiricism because none of its doctrines can be validated empirically.[3] But once empiricism is accepted as the equal partner of reason, the whole world—

and everything one has ever been told about it—becomes open to inquiry. If human reason and perception could penetrate the mystery of the planets, surely it could shed light on the political problems of inequality and injustice.

The Enlightenment

Thus it was that revolutionary thinking in science stimulated revolutionary thinking in social and political philosophy. Through the widely read writings of John Locke (1632–1704), the educated public was introduced to new ways of thinking about knowledge, government, and education. Sales of Locke's *Essay Concerning Human Understanding,* first published in 1690, shot up in 1703 when Oxford University banned undergraduates from reading it (Bowen, 1981). In this book, Locke advanced the empiricist's doctrine that "perception [is] the inlet of all [the] materials of knowledge." But "Sense and intuition reach but a very little way. The greatest part of our knowledge depends on deductions and intermediate ideas" (Locke, 1690a/1939).

Thirty years earlier, Thomas Hobbes had published *Leviathan,* a political treatise based on the beliefs that all human beings are descended from Adam and Eve and that originally they had lived in a state of nature in which all men were free and equal. This wasn't a paradise, however, because in a state of perfect freedom men would be continually at each other's throats, resulting in lives that were "solitary, poor, nasty, brutish, and short." They thus would contract away their freedom to the state in exchange for law and order. Hobbes strongly favored an absolute monarchy over democracy, arguing that a corrupt mon-arch might steal for his own family, whereas a corrupt legislature would steal for all the legislators' families.

In *Two Treatises on Government,* Locke also developed a rational theory of government on the Hobbesian concept of an original state of nature in which all men had "perfect freedom" and "executive power." Governments were formed by the contract of free men, as with Hobbes, but unlike Hobbes, men had the "liberty to separate themselves from their families and their government . . . and go and make distinct commonwealths and other governments as they thought fit" (1690b/1939). This natural liberty, which is really the innate human tendency toward rebellion (Chapter 5), isn't relinquished by agreeing to live under a particular government. Indeed, since all men have this liberty, all governments are, in Locke's view, contractual.

Already in 1620, the Pilgrim leaders had signed just such a contract on the Mayflower:

> We, whose names are underwritten . . . Do . . . covenant and combine our-
> selves together into a civil Body Politick, for our better Ordering and Preser-
> vation . . . ; And by Virtue hereof do enact . . . such just and equal Laws . . .
> as shall be thought most meet and convenient for the general Good of the
> Colony; unto which we promise all due Submission and Obedience.

By 1776, when Thomas Jefferson wrote in the Declaration of Independence that "Governments are instituted among Men, deriving their just Powers from the consent of the governed," the Lockian concept of government as a social contract to which men freely consented for their mutual benefit had become a well-established principle of democracy.

But there was never a time when all men had executive power, or when human life was solitary, because human beings have always lived in hierarchically structured groups. Nor was there ever a time when human life was significantly poorer, nastier, more brutish, or shorter than it was, for most people, in seventeenth century England. Hobbes was right about the existence of constant warfare among people "in a state of nature," but this was warfare between tribes and it didn't, in itself, lead to the organized state. The first cities were probably formed through the conquest of prosperous villages by nomadic tribesmen (Chapter 5). Locke admits that the victims of conquest haven't freely agreed to their new government, but he leaves their fate to God, and doesn't try to reconcile such nasty business with his theory.

The idea of human equality in Hobbes and Locke was based on their belief that all human beings were descended from Adam and Eve. Current genetic studies of living human beings around the world increasingly support the theory of a common African origin of *Homo sapiens* some 200,000 years ago, with a diaspora out of Africa some 110,000 to 140,000 years ago (Gibbons, 1995; Wilford, 1995). An alternative theory, based on the interpretation of a limited number of Asian fossils, holds that the different human races evolved separately from *Homo erectus* in different places, and converged to a common species through crossbreeding. However, the most telling evidence for a recent common origin is the similarity of peoples everywhere. All races can, and do, interbreed freely, producing healthy, fertile children. All people speak fully developed languages, which linguists believe are based on the same deep structure. All human beings are organized into hierarchical social groups and behave in recognizably similar ways. These facts are easily explained by a common African origin and a recent (100,000 to 200,000 years ago) separation, but are very problematic for a theory of converging evolution.[4]

In the eighteenth century, rational scientific thinking was replacing religious arguments among the more advanced thinkers—the *philosophes,* as they called themselves in France. The Western world was rapidly changing, both technologically and politically, and the aristocratic order, based on land and agriculture, was becoming ever more irrelevant as wealth and power accumulated in urban manufacturing centers. Education too, which at the highest level was the study of the Greek and Latin classics, was also becoming irrelevant. In the seventeenth century, all instruction at Harvard College was in Latin, and students were forbidden to speak English, even among themselves. Preparation for Harvard was provided by the Boston Latin Grammar School, which, battered but unbowed, continues to this day to offer a quality classical high-school education to qualifying Boston students.[5] But, by the end of the eighteenth century, an increasing number of English academies were being created, patterned after one established in Philadelphia by Benjamin Franklin. These institutions taught reading, writing, arithmetic, bookkeeping, navigation, surveying, and other technical subjects as they developed in the nineteenth century.

It was apparent that a growing industrial society needed ever more technically trained workers, and that economic growth provided opportunities for the social advancement of all elements of society. Thus education became linked with industrial growth on the one hand and social reform on the other. But industrial society is as hierarchical as feudal society, and this offends the very notion of equality that is the foundation of the Lockean social contract.

Progressivism

Around this contradiction there arose a radical literature and radical movements aimed not at the amelioration of social inequalities and injustices, but at the total transformation of society. Communism, the most familiar of these movements, aims at a classless communal society which is industrially competitive with capitalist societies. But before Marx, there was Rousseau, the godfather of all subsequent progressive attempts at radical social transformations.

Jean-Jacques Rousseau (1712–1778), the self-taught son of a watchmaker, was famous in his day for his radical views on society and education. He accepted the Lockean ideas of an original state of nature, of a social contract, and of the newborn mind as a *tabla rasa* or blank slate. But he saw nature and natural laws in moral terms, stating at the beginning of *Emile,* a novel describing the ideal education (1762/1979; Bowen, 1981): "Everything is good as it comes from the hands of the Maker of things; everything degenerates in the hands of man." In the original state of nature, men were free and equal; inequality and injustice arose as men formed governments and created classes. A utopian society, where all men and women are equal, would be based on the natural order, and the first step toward utopia is the radical restructuring of education. These themes—a romantic and reified view of nature, hatred of inequality and hierarchy, and belief that society can be radically restructured through education—are central to most progressive utopian ideologies current today. Rousseau, at the very height of the Enlightenment, set the postmodern anti-Enlightenment agenda.

Rousseau's idea of an education in accordance with nature was radical for an age in which infants up to one year of age were swaddled and older children were subjected to very coercive forms of discipline and instruction. Unbinding babies, letting young children learn by playing freely in their environment, and adjusting instruction in accordance with children's developmental states are elements of progressive Rousseauean education that have been accepted by almost everyone. Indeed, the modern secular educational system evolved from the nineteenth-century German effort to establish education on the foundation of a New Age holistic view of nature called *Naturphilosophie.* As with modern proponents of the same idea, *Naturphilosophie* sought a mystical connection between mind and nature, rebelling against the "Only the facts, Ma'am" attitude of empirical science.

But attempts to extend naturalistic learning beyond the preschool years has had only limited success. The most influential progressive schools were the *Kindergartens* that were modelled after the one founded by Friedrich Froebel (1782–1852) to inculcate *Naturphilosophie* notions about the unity and purposefulness of nature. Originally a complete schooling system, only the invaluable preschool part of Froebel's *Kindergarten* survives today (Bowen, 1981).

The problems of building a mass educational system on naturalistic, or constructivist, lines proved insurmountable, and by and large the world has adopted a modified version based on the work of the German educators Johann Friedrich Herbart and Wilhelm Rein. This is the classroom system of instruction through material aides—objects, charts, and diagrams—that was found to be necessary as the curriculum became too complex for "the constructive activity of the youthful mind . . . [by] itself [to] establish the connections between the manifest circles of ideas" (Rein, 1896; Bowen, 1981).

The moderate Herbartian system has been far more successful than the progressive ideology from which it evolved. Contemporary progressivism is, in fact, reactionary, since it reaches back to ideas and practices that failed over a century ago. It would be of little interest to us, except that it's very strong today among educators and academics, being the common denominator of constructivism, postmodernism, multiculturalism, radical feminism, ecoradicalism, and political correctness. Although a product of the Enlightenment, progressivism is fundamentally anti-Newtonian, antipositivist, and anticapitalist. It is moralistic, egalitarian, individualistic, and socialistic. It has been in the forefront of all efforts to expand the human rights and civil rights of blacks and women, while attacking the scientific and industrial systems that make it possible to meet the material, health care, and educational needs of everyone.

In education, progressivism is anti-authoritarian, advocating that the child learn primarily from experience, with a minimum of assistance from a mentor. In current terminology, teachers are to be facilitators, assisting, but not determining, their students' learning. This might be called the low-authority approach to education. The standard classroom could then be called the medium-authority approach, and the term "high authority" could be reserved for dogmatic schools committed to indoctrinating their students with a particular ideology or religion. This neutral terminology, although a bit awkward, may help clarify some of the issues that are raised by the need of modern societies to educate everyone to a reasonably high level in finite time with a finite number of finite teachers. It's also well to remember that the medium approach was developed for Prussian school children, who were raised in a culture a bit to the right of center on the rebellion–loyalty continuum.

Common School

Throughout history, governments have attempted to set educational policies, usually through unfunded mandates. Massachusetts, in the Act of 1647, required that every township "after the Lord hath increased their numbers to fifty households, shall then forthwith appoint one within the town to teach all such children as shall resort to him." But Massachusetts's educational policies were always ahead of the other states. Religious differences among the states prevented the adoption of a national educational policy at the Constitutional Convention of 1787, and only seven of the original thirteen states included education in their constitutions.

From 1820 to 1860, Boston grew from a town of 43,000 to a city of 178,000, largely through immigration from Ireland. And with the disestablishment of the Congregational Church in Massachusetts in 1833, the public school rapidly came to be seen as the major unifying force in a secularized democracy. Under the leadership of Horace Mann (1796–1859), Massachusetts developed the notion of the Common School, which would protect "society against the giant vices which now invade and torment it—against intemperance, avarice, war, slavery, bigotry, the woes of want and the wickedness of waste." "It may be an easy thing to make a republic," Mann once wrote, "but it's a very laborious thing to make Republicans." Following the Prussian model, Massachusetts developed the first graded schools in the United States, and in 1852, Massachusetts became the first state to mandate that every child between the ages of eight and fourteen receive at least three consecutive months of education a year (Fraser, 1977).

Compulsory education spread slowly to the other states, with Mississippi being, in 1916, the last state to require education for all (Bowen, 1981). The states of the Confederacy didn't begin adopting compulsory education until well after the 1896 Plessy v. Ferguson decision had decreed that segregation was constitutional as long as "separate but equal" facilities were provided for Negroes. Although this federal case was about transportation, the Supreme Court based its decision on an 1850 opinion of the Massachusetts Supreme Judicial Court that separate schools for Negroes were allowable if they were equal (Rushing, 1977).

Progressive ideas about the equality of man and his perfectibility through education motivated the early movement toward universal education, and immigration and the increasing urbanization of society forced the issue. In the past 150 years, state-supported education transformed the mass of the population of the industrialized world from rural agriculturalists—often serfs and slaves—into literate urban workers. This was a one-time event, which, through the resulting increase in the productivity of labor, played a major role in increasing living standards (Thurow, 1985). The small cost of teaching the son of a farmer enough reading and writing to be a productive factory worker was repaid manyfold by the increased value of the goods he could produce.

Since 1852, compulsory education in Massachusetts has gone from requiring sixty school days a year for six years to requiring 180 school days a year for ten years, a five-fold increase. Furthermore, since most jobs and training programs now require a high school degree, or equivalent, twelve years of schooling is the effective requirement. During this same period, a host of nonacademic programs—music, shop, sports—have been added to the curriculum, at an ever increasing cost. The limit to this escalation in government-supported education was reached in Massachusetts in the early 1980s, when, following California, it too put a cap on the rate at which property taxes could be increased. Whatever one might think of this, the political process probably did, in some crude way, measure the point at which the cost of state-supported education matched its economic benefit. Today, a K–12 education costs about $80,000 per student, or $1 per working hour over a forty-year work life. It's hard to see how this level of expenditure can increase much further, though we can hope that improved methods and technologies can increase educational productivity.

Integration

I was a senior at the University of Wisconsin in the spring of 1954, when the Supreme Court finally overturned its 1896 Plessy v. Ferguson decision. The case had been argued by Thurgood Marshall in late December of 1953, and the decision was awaited with much anticipation. During the interregnum, the campus chapter of the National Association for the Advancement of Colored People (NAACP) had a series of speakers who discussed the probable decision and its consequences. After these talks, I occasionally had coffee with some black students, where I learned a little about what life was like for them on a 99 percent white campus.

They had problems finding barbers to cut their hair, but no problem finding white women to date—the reverse of my problems. Our conversations were more about irony than anger, since we shared the conviction that after the madness of segregation was behind us, racial inequity and injustice would be eliminated. We truly believed that once black and white children went to school together, the major dividing differences between the races would vanish, as the major differences between immigrants and natives had been vanishing, more or less, for generations.

I had been raised to believe in the melting-pot model of America, and as an assimilated Jew I considered myself well melted. The notion was both an ideal and a description. The public school was to upgrade the lower classes, while teaching citizenship and democracy to all. In my kindergarten, we all sat in a big circle learning about love and humanity, just as Froebel had said we should.

My school district was all white during my early school years, but it did encompass all social classes, from the very wealthy to families on welfare. The wealthy lived in mansions and sent their children to boarding school or a neighborhood private academy; most Catholics went to the large neighborhood Catholic elementary school. In seventh grade, a boy from the private school enrolled for a year in my class to broaden his experience. His patrician dignity and good manners served him well, and he seemed to take the calls of "money bags" with good humor. His presence was the exception that informed us of the rule: there is an upper class, and it doesn't go to the common school.

Schooling in urban America in the 1940s was segregated not just by race, but by class and religion as well. Public education valiantly tries to stem the enclaving tendencies of an ethnically and socially diverse society, but it can't prevent it. When black families moved into the district after World War II, their children enrolled in Shakespeare Elementary School without incident. A few became "walk-to-school" buddies of mine. But the school couldn't prevent the transformation of the district from all white to all black in the course of a decade.

Seven years after the Civil War, Frederick Douglass wrote an article in the *New National Era* that cogently stated the case for racially integrated schooling:

> Educate the poor white children and the colored children together; let them grow up to know that color makes no difference as to the rights of a man; that both the black man and the white man are at home; that the country is as much the country of one as of the other; and that both together must make it a valuable country. (Meltzer, 1965)

Douglass made clear in the article his belief that the common class interests of poor whites and blacks should transcend their racial differences. He didn't expect poor blacks (or poor whites) to go to school with middle-class whites. Indeed, in his day, the struggle to provide free education for all whites was strongly resisted by conservative forces, who saw no reason why the rich should pay to educate the poor.[6]

In May 24, 1954, Chief Justice Earl Warren read the Supreme Court's unanimous decision in Brown v. Board of Education:

> To separate [Negro children] from others of similar age and qualifications solely because of their race generates a feeling of inferiority as to their status in the community that may affect their hearts and minds in a way unlikely ever to be undone.... We conclude that in the field of public education the doctrine of "separate but equal" has no place. Separate educational facilities are inherently unequal. (*Time,* 1994)

Today, through digital magic, I can call up the stories and pictures of those eventful days on my computer. Particularly dramatic was the integration of Central High School in Little Rock in 1958, which required federal troops to protect nine black students from hundreds of defiant whites. These students had been specially prepared for their ordeal, and maintained great dignity in the face of the outrageous threats and abuse hurled at them from the large angry crowd that had surrounded the school. They later told reporters that they had received nothing but friendly treatment from the teachers and students in the schools, which was just what an anxious country wanted to hear.

Court rulings progressed from outlawing enforced racial separation in the South, to demanding racial balance in the North (Armor, 1995). In 1974, the Federal Court put into effect a Draconian plan to integrate the Boston Public Schools by assigning half of the 94,000 students to schools they would not otherwise have attended. To accomplish this, 18,000 students (half of them white) were bused back and forth across the city. The entire liberal establishment was behind this plan, which clearly violated the integrity of the city's ethnic enclaves by depriving them of their neighborhood schools. This was considered a small price to pay for the great racial harmony that was to follow as students of different races and classes learned to get along with one another.

Overlooked by everyone was the tenacity with which blacks would cling to their identity. Unlike immigrants, most black students had no great interest in making friends outside their community, let alone learning white ways. In a nightmare of social engineering, positive attitudes toward school came to be viewed by some blacks as "white," and so negative attitudes were encouraged to reinforce "blackness." Yet the drive to integrate continued relentlessly, with attacks on any programs, such as tracking and ability grouping, that might produce classes with different racial ratios. The result was as predicted: massive defection of whites, and some blacks, from the Boston Public Schools. Today, the system enrolls only two-thirds as many students as it did before integration.

In 1989, Boston's court-ordered integration was relaxed. Under a modified plan, parents can now pick the schools they want their children to attend (Gold, 1989). Choices are honored as long as they don't lead to racially imbalanced schools. This seems to be working, and the system has stabilized at 18 percent white. Other cities haven't been as fortunate. The Hartford system is 92 percent minority, and in a 1995 decision, the court ruled that integration with suburban schools was not required to remedy this. Thus forty years of court involvement in integration has failed to achieve its objectives. Perhaps it's time to consider a broader range of options for providing a common education for the common good.

Conclusion

In spite of the long history of education, education in the industrialized countries is dominated by the rise of mass education in the last 150 years. And mass education has been dominated by a particular educational strategy that developed in a particular culture. Just as China, in the third century B.C., selected Confucianism from among many competing systems as its official educational and philosophical system, so the West, more than 2,000 years later, selected the graded Prussian system from among many other modes of instruction. In both cases, systems that had been developed by reformers to restructure society were adopted by society to maintain the social order.

There is an inevitable contradiction between the progressive doctrines of educators—with their emphases on naturalism and individualism—and the practical needs of schools, businesses, and society as a whole for discipline and order. This is especially true in the English-speaking countries, where a long tradition of personal freedom conflicts with the coercive nature of compulsory education. And no matter how enlightened the pedagogy, there is nothing natural about a classroom. All schooling can't be modelled on kindergarten.

Part of the mission of the common school is to develop a common national identity by limiting people's natural enclaving tendencies. The present system of medium-authoritative education succeeds only with children who come from families and cultures that enforce a minimal level of discipline and obedience. The excessive freedom of the West makes it difficult for the undisciplined to successfully navigate around the many seductive distractions and deadly temptations that the society permits. Obedient Confucian societies, which tolerate fewer behavioral options, are better adapted to Western educational methods than is the rebellious West itself.

The failure of the common school in the United States to educate millions of students to the level they need to survive in an industrial and commercial society is a matter of grave national concern. The role that discipline plays in this failure is receiving increasing attention. The American Federation of Teachers has called for imposing sanctions—expulsions and flunking—to enforce stricter discipline and higher academic standards (1995). Albert Shanker, long-time president of the AFT, has even endorsed grouping students by ability (often confused with tracking) to provide a better educational experience for all. This position, long opposed by progressives because of its negative affect on integration, is being reexamined by some black intellectuals (Raspberry, 1995). This issue is discussed further in Chapters 9 and 10.

The more complex and technical society becomes, the more important education is in determining one's place in the hierarchy. Literacy is still a major determinant of social position, and the computer is adding a whole new layer of complexity on top of that. Will the computer lock out even more people from the "mainstream," or will it, like power machinery, increase the productivity of all segments of society?

The next chapter examines the perplexing problem of computers in education. Are they the ultimate learning machines that will enable students to break through to new intellectual heights? Or are they a counterproductive luxury that drains time and money away from more direct learning experiences? Must schools teach "computers," and if so, what exactly should they teach? Will today's lesson be relevant tomorrow? I've been personally involved with using computers to teach science since 1980, and I've been on all sides of the issue, falling in and out of love with the educational role of computers with great regularity.

7

Of Chalk and Chips

After printing, the most important technological innovations in education have been inexpensive paper and the blackboard. We take it for granted that every child has unlimited amounts of paper on which to write and draw, but this is really a luxury of industrial societies that isn't available to many children in undeveloped regions of the world. In ancient Greece, children learned to write by inscribing their letters on wax-covered boards, and small slate boards are still used by the poor in India and elsewhere. Both are crude, but reusable, media.

The blackboard was new enough in 1826 for the American physicist Joseph Henry to record seeing one in the chemistry lecture room at the U.S. Military Academy (Henry, 1972).[1] To this day, the blackboard is a universal feature of every classroom from the remote New Guinea highlands to the halls of ivy. Only in the past decade have new technologies and styles of teaching begun to supplant chalk as the master medium of instruction. The humble blackboard is the epitome of a successful educational technology. Its essential characteristics of universality, accessibility, and flexibility become apparent, as do the functions of a gene, only when they are lost.

In 1980, I replaced some blackboards in the physics laboratory with new-fangled whiteboards that use marking pens instead of chalk. The purpose was to eliminate chalk dust that might harm the new-fangled computers I was introducing. The trouble is that the marking pens dry out very rapidly if they aren't covered when not in use, so they are almost always dried out. Half the time, when I reach for a pen to write something on a whiteboard, the pen is dry and I have to scramble to find a new one. Since the pens are expensive, extras are locked up in a cabinet somewhere. This isn't a minor inconvenience, since one of the great advantages of the chalkboard is that it's always available for instant use. There is always some chalk sitting unmolested in the chalk tray.

My problem is a peculiarity of my environment. New technologies that work wonderfully in one environment—say a corporate conference room with a staff to check that there is always an adequate supply of marking pens—can fail in schools where there is little money, staff, or time for maintenance. And though technology has vastly increased access to and control over information for those who know how to use it, it may have limited impact on students who can't properly use the index in a book. Teaching these students to use a spelling checker on a computer may result in their never learning to use a dictionary. The effectiveness of technology for education is always limited by the weakest links in the whole implementation process. You can't connect a drinking straw to a fire hydrant.

Yet the adoption of microcomputers by U.S. schools has been explosive, going from essentially zero in 1980 to better than one for every nineteen students by the early 1990s. The computer's educational roles change from year to year, as their functionality evolves, and today their purposes are as unclear as they are unquestioned. What is clear, is that whatever their purposes today, they aren't what I, or any of the computer gurus of the early 1980s, thought they should be (Bork, 1981; Cromer, 1981; Pappert, 1980).

Calculator

In 1960, the most advanced calculating tool at the Harvard University Cyclotron Laboratory was the Marchant calculator, a marvel of mechanical ingenuity with eighty keys, arranged in eight ten-key columns. To enter 7.85 you pressed "5" in column 1, "8" in column two, and "7" in column three. This set levers, which, when the enter key was pressed, advanced gears that turned circular dials. The Marchant easily added and subtracted, and with much huffing, puffing, and shifting of its carriage, it could multiply and divide as well.

The Marchant may have had the most moving parts of any mechanical device of its size, and was clearly near some limit of mechanical complexity. In 1961, I switched my research computing from the Marchant to the IBM 704 computer that had recently been installed at MIT. Codes were mechanically punched onto cards and physically transported to the computer center, where in the goodness of time they would be processed and returned to me with output indicating some of my many errors. (Only when all my syntactical errors were corrected would the computer even consider my logical errors.)

Throughout the 1960s and most of the 1970s, engineering and science students continued to use slide rules for their day-to-day calculations, as did their professors. The controversy then was whether students should be taught how to use the slide rule or should be expected to learn it on their own. Slide rules were considered too expensive and too difficult to use for nonengineers, who used trigonometric tables and long multiplication and division to do the problems in their physics courses.

For me, the age of technology in education began in 1972 when the first electronic calculators came on the market. A four-function (addition, subtraction, multiplication, and division) machine, about the size of a small book, cost $200. Two were purchased for the physics laboratory at Northeastern and, after being carefully bolted down, were made available to the students during their laboratory period.

Two years later the Texas Instruments and Hewlett-Packard slide-rule calculators came out for about the same money, and these were added to the laboratory. These calculators, now the standard scientific calculator, had all the trigonometric and logarithmic functions of a slide rule. As more complex calculators kept being developed, with graphical and programmable capabilities, the price of the scientific calculator dropped, so that soon every student, engineering and nonengineering, had one. By the late 1970s, the scientific calculator had replaced the slide rule and had equalized the computational power of engineering and nonengineering students.

This rapid replacement of one technology by another required two factors. First, the new technology had to have at least the same capability as the old, and second, the price of the new technology had to drop below the price of the old. When I was a student in the 1950s, the price of a slide rule was five to ten times the price of a physics textbook, whereas today the price of a scientific calculator is less than one-fifth the price of a physics textbook. The new technology required almost no change in the teaching practices of the science faculty, except that professors could assign problems with more realistic numbers.

The stabilization of education practices at the slide-rule level is a long-standing frustration of mine. I certainly thought that computers would replace the calculator, just as calculators had replaced the slide rule, and that college physics would be dramatically transformed (Cromer, 1981). But, in spite of much effort by many capable people, this hasn't yet happened. Why? The usual reason given is the conservatism of college physics teachers, a notoriously conservative group of rascals. But this doesn't go deep enough. We did, after all, accept the calculator, in spite of the obvious superiority of a well-adjusted bamboo slide rule.

To understand the opportunities and limitations of technology in education, I offer my own experiences as a case study.

Apple II+

In 1980, with funding from the National Science Foundation and Northeastern University, I purchased ten Apple II+ computers for the introductory physic laboratory. At about $2,000 each, they had 0.048 megabytes of random-access memory, a floppy-disk drive that read disks with 0.12-megabyte capacity, and a 1-MHz processor. (Today, for half the price, you get ten to 100 times these capacities, plus 250 megabytes of hard-disk memory and a CD-ROM player.) The Apple II+ came with the BASIC programming language built in, game paddles, low-resolution color graphics, high-resolution monotone graphics, and sound—everything I needed for developing simulated experiments.

In a typical simulation, students set the initial speed and angle of a projectile by turning the dials on the game paddles. Then, when they pressed a button, a dot moved across the screen the way a real projectile would move through the air. Students measured the time and distance of the dot using a stopwatch and ruler, and analyzed their data as they would a real experiment. The novel feature of these simulations was that the students made real measurements of the computer display. In all, I developed six such experiments, which were used for four or five years in our laboratories (Cromer, 1980).

At the same time, I started a small software publishing company, EduTech, to market programs that I and other teacher-programmers had written for the Apple II+. The Apple II+ served us well, providing simplicity, versatility, and universality. All Apples were the same. As late as 1990, I could take an EduTech disk written in 1982 and have it run on an Apple II+ or IIe in Indonesia. There were no problems with graphic cards and a hundred different system configurations that IBM gave us. This was important because our programming resources were limited to what we could do ourselves, and our market was limited to high-school and college science classes. Once this market got divided up among several computers it became too small and too technically challenging for the amateur to reach. Nevertheless, by the time I transferred ownership of EduTech to one of my authors in 1985, the company had sold my works, and that of about a half dozen other authors, to several thousand schools and colleges worldwide, and I just about broke even.

I had envisioned the computer becoming a highly versatile part of the physics laboratory in high school and university, and to a large extent it has.[2] But I had also hoped that students would start to solve more complex physics problems with the computer. For example, the motion of a planet around the sun is determined by a pair of coupled differential equations, the solution of which is generally done in graduate or advanced undergraduate physics courses. But using simple numerical techniques, even high school students can write programs to solve such a problem. The microcomputer is particularly suited for this, because they can be programmed to plot the motion of the planet as it is being calculated.

In 1980, I began doing this sort of thing on the Apple II+, and to my surprise and frustration I found that the simplest numerical technique, called the Euler approximation, didn't work. Instead of drawing a closed elliptical orbit, as predicted by theory, the numerical solution gave a very unacceptable spiral. It just so happened, as I was trying to figure out what to do next, Abbey Aspel, the daughter of a friend, had been assigned the same problem by her high school physics teacher. At the time, Newton North High School was using a minicomputer that only gave numerical printouts. Aspel had to plot her orbits by hand, and found, as I had, that they spiralled outward instead of closing on themselves.

She discussed her problem with me, but was dissatisfied with my answer that the fault lie in the approximation itself. She believed she had made a programming error and decided to reverse the order of two lines in her code. Having been working on the same problem, I assured her that her original code was correct and that her proposed change was absurd.[3] She went ahead with her change anyway.

I was a bit miffed by her stubbornness, feeling that she should have had more respect for my superior knowledge of numerical methods in physics. I prepared a short speech on the matter for the next time I saw her—something about the deference due to authority. But when I arrived at her house, she was engrossed in plotting perfectly closed ellipses. How had these been calculated? From her absurd modification, of course. I canceled my speech, and returned quickly to my Apple where I instantly made her modification to my code and watched in dumb-founded amazement as the dots on the screen traced out a closed elliptical orbit.

To try to understand what was happening, I looked into books on numerical methods. These rejected the Euler approximation as too weak to be of any practical value, and developed stronger, but far more complex, numerical methods. There was nothing in these books about improving the Euler approximation by simply reversing the order in which one calculated things. After months of agonizing over this, I finally hit upon a somewhat cumbersome proof of why this reversal worked. In the Euler approximation, the overall error in the approximation increases as each point is calculated, whereas in the modified ap-proximation, the overall error oscillates, repeatedly passing through zero (Cromer, 1981).

Computers in Physics

Thus, by 1981 we had the technology and the methodology for revolutionizing the way we taught introductory physics. And in the second half of the decade, when computers became generally available to students, a number of numerical-methods courses were developed (Gould and Tobochnik, 1996; De Jong, 1991). But these never affected the standard courses, which, judging by the major textbooks in the field, have remained in the slide-rule era.

Changes in teaching are difficult to introduce at a large research university, because the large-enrollment courses often are taught by a number of faculty of varying levels of commitment. Young faculty can't afford to get entangled in time-consuming innovations that could distract them from their research, and older faculty aren't generally interested in someone else's innovation. I was able to try teaching physics through a computational approach only when I had my own course, either the one section of honors engineering or the one section of Physics 1 offered out of sequence. From my experience, I found some conflicts between technology and education.

At the time I taught the honors engineering course (1986–1988), the students were simultaneously taking a course in PASCAL, then a popular programming language. Thus my approach was in conformity with their overall program. As one student remarked, "I've been studying programming since tenth grade, and this is the first time anyone asked me to do anything with it."

My approach required the students to program the computer to calculate and plot orbits and to print the graphs. Although these capabilities have always been standard on the Macintosh, IBM systems require special hardware and software for putting graphics on a screen and obtaining a printout. The first year went well: there were systems around the university that could do what I wanted and consultants who knew the specific commands that had to be inserted into each program to make them graph and print. But by the second year, there had been major changes in hardware and software configurations of these computers, and I could find no one who knew how to get them to do what I needed.[4]

The situation would have been very different if the university had required every freshman engineering student to buy a specific computer with a specific software and hardware bundle. This would have allowed the faculty to learn a common system and to develop courses around it. Instead, most universities opted to make computers available to all students on a first-come-first-served basis at centrally administered access stations scattered around campus. This put the universities on an endless technological treadmill, investing millions of dollars a year just to stay in one place, while creating a jumble of systems of different manufacture and vintage.

Technology can rapidly become so complex that it defeats its own purpose. In 1993, my department spent $15,000 on a multimedia computer system, with laser-disc player, CD-ROM player, liquid-crystal display panel, and special software (Podium) that allows one to write a presentation that intermixes segments from computer programs, laser disks, and CD-ROMs. It's absolutely marvelous. But it takes hours to write a single lecture, and to present the lecture one must assemble at one point the computer, the laser-disc player, the CD-ROM player, an overhead projector, and the liquid-crystal display panel, as well as the software, CD-ROMs, and laser disks pertinent to the day's lecture. To be practical, all the equipment must be permanently connected and dedicated to this one purpose. At present, the system is used several times a week by the few faculty who are devoted to it, with no evidence that it has any effect on learning that couldn't be achieved with a few live demonstrations.

Universality

Education at present, and probably into the indefinite future, requires more flexibility than is possible with such complex systems. To be effective in education, a technology must be universally available. The blackboard is still the best example of this: no classroom is without one, and they work the same way all over the world. The overhead projector competes effectively with the blackboard in many instances, and is destined to replace it. The great advantage of the overhead projector is that it can display both spontaneous writing and prepared transparencies, as well as a variety of other instruments, such as electrical meters, that are designed specially for overhead projection.

The personal computer is as revolutionary an innovation as paper, but its full impact on education won't be realized until it becomes as universal as paper, that is, until every child can put her hands on one any time she wants. In spite of the millions of computers in schools, schoolwork is still done almost exclusively by hand. A school child still learns 104 alphabetic characters—twenty-six upper and lower case letters in both block and cursive form–even though the computer has already made cursive writing obsolete for most adults. At some point, children will have to spend more time practicing their touch typing than they do their penmanship.

In the early 1980s, I had expected the price of computers to fall as had the price of calculators. Instead, computer manufacturers kept their prices roughly constant while dramatically increasing the functionality of their product. In the process, they discontinued their older models, instead of selling them for a lower price. The calculator manufacturers, on the other hand, reduced the prices of their basic and scientific calculators as they marketed more expensive programmable and graphics calculators. Had the computer manufacturers done the same, we would have the equivalents of an Apple II+ or Mac II selling for under $200.

The computer revolution in education would be greatly accelerated if the manufacturers agreed to produce a standard school computer with standard software that sold for under $200. At this price every elementary-school student and teacher could have one, transforming the school environment from paper to keyboard. This point may soon be reached. Already some laptops are selling for under $1,000, and for $1,600 a family can buy a complete multimedia computer system, with CD player, color monitor, black and white printer, fax modem, and gigabytes of software including dictionaries, encyclopedias, atlases, and almanacs.

Schools can never keep up with the explosive pace of the microcomputer revolution, since it takes decades to develop effective computer-based curricula and to train every teacher in it, whereas computers change every two years. Schools will never have the time and money to stay at the cutting edge of technology, nor is there any reason that they should. Yet poor school districts complain if they can't upgrade their computers to run the latest software. Clearly, cutting-edge and universality are incompatible objectives, since cutting-edge is always scarce and expensive.

But is universality good for everyone? Perhaps there may be some cold-hearted logic in having cheap old-fashioned computers for poor children, but why should a rich suburban school system forgo the latest technology? Shouldn't it provide the best for its students?

Indeed it should, but I'm arguing that a universal computer is the best for everyone's children because, over time, every teacher will incorporate it into her teaching. It's of little value to a rich school to have a lot of new equipment that only a few teachers have the time or interest to fully use. The technology doesn't do much good on its own. It must be part of a carefully developed curriculum taught by trained teachers. The time scale for this development and training is decades, not years.[5]

Research in Educational Technology

Research in the educational use of cutting-edge technology is often distorted by the funds that support it. With millions of dollars for new equipment and development, researchers can quickly evolve projects that bear little relationship to the realities of a typical classroom. My own work in 1980 was supported by a grant of laboratory equipment. The computers I bought had to be put on line quickly, so they couldn't be too complex or esoteric. I decided on the Apple II+ because, even then, it was emerging as a consumer product that I could program myself. I was running a nine-hundred-student laboratory at the time, so I couldn't take the risk of buying a machine that was beyond my own expertise. By staying within my own limits, I was able to develop software that was of use to my own students and to students in a thousand other classrooms around the world.

At the same, the Mathematics Department at Northeastern received a $500,000 grant to develop software to teach mathematics. This large grant included money for state-of-the-art microcomputers and for computer analysts to program them. The investigators weren't interested in the $2,000 Apple; they were interested in the educational computer of the future, the $5,000 Terak. No matter that the Terak came without a suitable operating system—they had hundreds of thousands of federal dollars to develop one. No matter that the rest of the educational world was going Apple—the purpose of the grant was to go where no one had gone before. And so they bought five Teraks—perhaps the only five ever sold—and hired programmers to write a usable operating system and authoring language. With these they developed some very fine calculus modules that were used by students at Northeastern for many years, but which, since no other school used Teraks, had no further influence.

Any research project is necessarily risky, and $500,000 isn't a lot of money to invest in checking out the potential of a new technology. The Terak wasn't an unreasonable choice. A consortium of educators, centered at the University of Utah, had endorsed it because it had greater graphical resolution and processing power than the Apple. But at the time, education didn't need more resolution and power. Although the II+ didn't have lower case letters, and so was intrinsically unsuitable for word processing, software came out that turned it into a true Wysiwyg ("What you see is what you get") word processor. It was amazing. The software formed the characters in the computer's graphics mode, much as the Macintosh does today, and each page was printed as graphics on an Imagewriter printer. This gave the II+ the capability of "true descenders"— p's and g's that descend below the line—before they were commonly available on character printers. For many years, education was well served by fully exploiting what it had and understood.

Research in educational technology often looks like a solution in search of a problem. It's easy to fall in love with high-tech gizmos—I've done it myself— and to lose sight of the ultimate objective: the education of a student. There is often a fine line between a technology that educates a student and one that entertains its developer. Furthermore, even when the technology is of unquestionable educational value, it may be replacing, at great expense, less expensive and equally effective methods. After all, Isaac Newton did very simple and convincing experiments to demonstrate his Third Law of Motion without the need of force probes and microcomputers.

Funding agencies have come to equate technology with the complex and the expensive, thus promoting research in that which is, by definition, unsuitable for mass utilization in schools. To some extent, the funding and promoting of expensive technology overshadows the many ingenious devices developed by inventive teachers using universally available materials (Cromer and Zahopoulos, 1994; McGervey, 1995; Morse, 1992; Edge, 1987) For example, Fig. 7-1 shows a simple lung model built from a plastic soda bottle. In a way, this is a very high-tech device, since the plastic soda bottle is a product of advanced polymer chemistry. The availability of a bottle that can be squeezed, cut, and pierced, makes it suitable for dozens of demonstrations and experiments that can't be done with glass (Ingram et al., 1993; Salzsieder, 1995).

Quiet Revolution

Although computers haven't yet become as cheap as calculators, calculators have, for school computing purposes, become as functional as computers. By achieving universality, these calculators could revolutionize the way we teach mathematics and science. At the very least, they will shake up the stupor of university education, since students are coming into college classrooms with more calculating power than their professors.

Graphics calculators manufactured by Texas Instruments, Hewlett-Packard, and Casio sell for under $100 and have roughly the computing power of an Apple II+. The TI-82 has 32,000 bytes of memory, enough to store scores of programs, equations, graphs, and tables. Its small liquid-crystal screen is ninety-six pixels wide and sixty-four pixels high, and holds eight lines of text . Its many built-in routines make it easy to plot equations and data and to calculate regression lines and correlation coefficients. An auxiliary unit, the Calculator Based Laboratory (CBL), records voltage, temperature, distance, and other physical data and sends it to the calculator through a connecting cable. The calculator instantly graphs and analyzes these data and can send its graphs to a microcomputer, from which it can be printed on a standard printer.

The graphics calculator is making rapid inroads into the teaching of mathematics in high school. They are required of students in Advanced Placement calculus courses, and must be used in taking the AP test. In Boston, mathematics professors from Northeastern are teaching basic algebra to eighth-grade students with graphics calculators that are handed out each day in class. The students learn to draw "stars" by plotting equations of the form $y = mx$ with different values of m. Workshops are sprouting up everywhere to acquaint teachers with the exciting potential of this technology. Even cheaper graphics calculators are being marketed to middle-school students, so, for the price of a pair of sneakers, every student may soon own one.

University education is definitely behind on this one. I first learned of the graphics calculator in late 1994 and gave a colloquium on them to my department in early 1995. None of the physics faculty knew of them, even though

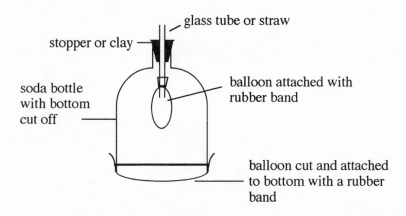

Figure 7-1 A model of the lungs made from a 1-liter soda bottle. The bottom half of the soda bottle is cut off. A balloon, representing one lung, is attached with a rubber band to the end of a glass tube or straw, representing the trachea. The balloon is placed inside the bottle, representing the chest cavity, and the tube passes through a one-hole stopper sealing the neck of the bottle, or through a seal made from clay. A sheet of rubber cut from a balloon is sealed across the bottom of the bottle with another rubber band. This sheet represents the diaphragm. When it is pinched in the middle and pulled down, the volume of the chest cavity is increased. Air flows through the tube into the balloon, and the balloon expands. When the diaphragm is released, the volume of the chest cavity decreases. Air flows out of the tube, and the balloon shrinks.

they are required in some college calculus courses and virtually all engineering students had them. A graphics calculator can easily store all the equations used in a physics course, enabling students to have crib sheets on all their tests without their physics professors being aware of it. So the first reaction of the faculty to this startling revelation was to ban graphics calculators from examinations. But physics professors can't hold out forever against a powerful technology that's already being heavily utilized in high school and college mathematics courses.

Perhaps the graphics calculator is the universal calculating machine of my dreams. Perhaps it will finally extricate physics education from its hopelessly outdated curriculum. With each student in a classroom having in his hand the power to solve and plot coupled differential equations, traditional problem-solving techniques becomes questionable.

For example, instead of using the formula for the solution of a quadratic equation to find the time a projectile travels a certain distance, my students are taught to quickly plot the distance the projectile traveled for all times, and to read the particular time for a particular distance from the curve. The advantage of this approach is that it's very general, given the extraordinary fire-power of a graphics calculator. Virtually any mathematical equation can be plotted, eliminating most of the algebraic manipulations traditionally needed to solve physics problems. Is it worth spending less time on algebra and more time interacting with the technology? I believe it is, because the technology radically increases the range of problems that can be solved. With the physics equations safely stored in their calculators, students can now concentrate on learning to effectively use the enormous computing power of these machines. Many more students are likely to become creative problem solvers this way than ever mastered the technicalities of algebra and calculus.

Because students always have their graphics calculators with them, they can practice their programming in class. The justification for this is that the process of programming the calculator to solve a problem focuses on exactly what is needed for a particular task—nothing more or less. The process, in effect, programs the students. The instructor's programs and plots can be projected on a screen, using a liquid-crystal display panel placed on an overhead projector.

It will be decades before any substantial number of physics courses are transformed in this way. I have only started the process myself, and I'm alone in this in my department. The textbooks are still written for the slide rule. The biggest disincentive for change is the continued introduction of newer models of calculators, before even the pioneers have learned how to use the existing models. I'm already confronting two models in my classes. Schools will have to insist that all students use the same model if they expect their instructors to cope with this complex, but enormously fascinating, technology.

Future of Technology

A friend of mine, who once worked for Pepsico, told me that the big idea there was "stomach share." The average American drinks so much liquid a day, and Pepsico's objective was to increase its share of the liquid that went into the average stomach. Just so, technology's objective is to increase its share of what goes into the average mind.

It's a zero-sum game. Time programming a calculator is time not listening to a physics lecture. Time on the Internet is time spent not reading a book. It would be nice if schools that are spending millions of dollars for new computers had done a cost-benefit analysis showing that time spent on computers was more productive than time spent on more traditional activities, but I know of no such studies.

There are two distinct reasons for computers in the schools. One is to teach traditional subjects better, and the other is to teach specific computer skills—programming, desktop publishing, graphical rendering, and so on. A composition teacher might insist that her students do their work on a computer so that they can quickly revise it after a first reading. For short assignments, the students might work for a period or two in the computer lab, but for longer work they must have at-home access to a computer. The basic word-processing knowledge for such work is trivial and will be picked up readily by students in the course of their assignments, provided the software isn't overly complex and isn't changed with every equipment upgrade.

The teaching of more sophisticated software applications is very different. The knowledge and equipment needed to work with photographic-editing software or three-dimensional rendering software is beyond most schools. A few specialized schools may be able to offer state-of-the-art instruction in morphing, but the average school can't. Schools that try to keep up with the latest in software, scanners, gigabyte hard drives, and work stations will, like the Soviet Union, exhaust themselves in a technology race they can't win.

Far more important than schools teaching some students sophisticated computer applications is that all students have full-time access to a computer with which they, and all their teachers, are comfortable. This equity issue will be solved only when schools require all students to buy a standardized low-cost computer. Public schools have been reluctant to require parents to pay any extra school expenses. But this could change. Payment plans of as little as five dollars a week could be made available to poor families, who would be proud to be contributing in such a tangible way to their children's education.

Up to now, the United States has been unable to standardize anything in education, making the delivery of educational services a nightmare of inefficiency (Chapter 10). If there is ever to be a standard computer, it will have to be specified by some consortium of computer manufacturers. The idea would be to use generic components, or older proprietary chips that could be licensed at low cost, in a computer designed to meet the specific needs of schools. The design would be determined by talking to teachers, a strategy that, in their arrogance, high-tech companies haven't always done.

The classic case of "not knowing its market" is the Plato system, which lost Control Data Corporation hundreds of millions of dollars in the early 1980s. At an educational convention back then, a CDC salesman showed me the Plato physics course that ran on a dedicated CDC computer. The student worked through the course frame by frame, answering problems as they were encountered. "Suppose," I said "that a student needs some time to think about a problem. Can he log off and return later to the same point in the program?" Since physics problems can take a long time to solve, this feature would allow another student to use the computer while the first one was working off-line.

But Plato didn't have this feature. Apparently the student had to work through the lesson to some fixed end point before returning to a menu. For whatever time it took him to solve a problem, the expensive computer sat idle.

I was intrigued by this inefficiency, because EduTech had published some programs written by James Friedland, a high school physics teacher, that interacted with a whole class of students. Friedland's problem was that he had one Apple II+ computer for thirty students. To use it effectively, he wrote programs that allowed each student to log on by name, receive a problem, and then go off-line to solve it. In his program, TARGET, the problems involved calculating the initial speed and angle of a missile in order to hit a given target. When a student returned to the computer, she typed in her name and was returned to the situation she had been given. She then typed in her solution and the program animated the track the missile would follow. If she did her calculation correctly, she was rewarded with a hit and a new problem. The program even kept track of how many problems each student did and how many they each got right.

Only teachers and students can know some of the special hardware and software needs of a standardized middle-school computer. Certainly the computer would have to be rugged enough to be carried around for years in a backpack. A hard drive might be inadvisable, since a crash could ruin a whole term's work. For lightness and low cost, it wouldn't have batteries, but would run from powerpacks at home and at each desk in school. The operating system and application software would be small enough to fit on a floppy disk, with a lot of room left over to store documents. A simplified works program would probably meet 99 percent of a student's need through eighth grade and beyond.

Most important, the basic design of the computer and its software bundle would remain fixed for at least five years. This would give time for most teachers to learn to use them and to incorporate them into their curricula. The educational system is immense, and it learns slower than its students. A moratorium on change would also give time for third-party vendors to develop software and hardware that could greatly enhance the value of the machine.

Will a low-cost standard computer ever be manufactured? Not as long as schools are willing to spend millions of dollars for machines that are fancier than their students need. But budget realities, if not sanity, may someday cause schools to reassess their priorities. If schools were to radically de-escalate their technology race, the sale of two-hundred-dollar-computers to four million fifth graders a year might start to appeal to the computer manufacturers. It already is appealing to the calculator companies.

To change education, computers must be as integral to the process as a pen. And for most school purposes, a Bic is as good as a Cross.

Both computers and graphics calculators are menu-driven, making them easy to learn and "intuitive" to use. Yet to the raw beginner, the nested sequences of options at first looks like an impenetrable maze of choices. Only after running through this maze a number of times does the process become intuitive. The term "maze" isn't just a metaphor, since a mouse running through an experimenter's maze is also confronted with a sequence of options to be learned. How does the mouse do this? Is there any connection with how humans learn? If so, does this similarity tell us anything about effective teaching? The next chapter investigates these questions using a computer model of maze learning. The purposes are to illustrate how computers are used to model complex processes and to obtain some insights into certain aspects of the learning process itself. In particular, we will use the model to study the competing needs for efficiency and flexibility in a self-organizing system, and to disentangle the effects of nature and nurture on the intelligence of cyber mice.

8

Of Mice and Men

Many years ago, while on sabbatical in England, I became obsessed with the problem of writing a computer program that could learn. Learn what? All I knew of learning was that psychologists studied it by running mice through mazes and that, as a result of a reward a mouse got for finding the goal, it ran the maze faster on its next trial. The problem, then, was to write a program that would model the behavior of a mouse in a maze.

Since then, computer modeling of complex systems has taken its place, alongside experimentation and mathematical theory, as a third method of scientific investigation. The modeler acts first as a theorist, simplifying the system of interest—the weather, the brain, or the mouse—to a few elements and conjuring up mathematical rules for how these elements interact with one another. He then acts as an experimentalist, studying with a computer how the rules affect the behavior of the elements. The purpose of the work may be to develop a realistic model that can, from inputted data, predict (say) the weather several days in advance. More often, however, the purpose is to get some insight into the interplay of the elements. For example, programs that model how the brain learns to recognize complex patterns are having a profound influence on how we think about thinking (Churchland, 1995). In such cases, the model may be interesting even if it doesn't predict anything real, although it's always more interesting when it does.

In the familiar pencil maze, one looks down on a visually complex pattern of pathways and deadends. The problem for a mouse inside a maze is very different, since it sees only one turn or opening at a time. It can't draw a map of the maze, but must discover and remember the particular sequence of movements that gets it to the goal in the shortest time or with the least number of unnecessary moves. From this point of view, the learning of a maze "from the inside" has close similarities to the learning of any sequence of discrete decisions, such as the moves in a computer adventure game, or the operations of a menu-driven computer program or graphics calculator.

I had the delightful opportunity in England to test my model by walking through the famous maze at Hampton Court, which is formed of ten-foot-tall hedges. I got just as confused on my first trial as my model said I should have, but I never found out if I would have improved as much on a second trial as the model predicted. Many amusement parks now have mazes, allowing human beings to experience for fun what mice have had to do for work.

The most interesting aspect of the model is that it shows how organized behavior at one level emerges from a matrix of disorganized behavior at a lower level. Because of the frequent back-tracing that occurs when running a maze for the first time, there are a vast number of paths of various lengths through a maze, yet after a few trails the model finds the shortest (or nearly shortest) path, as do real mice. To some students, a word problem in physics or mathematics is like a maze to a mouse, because they can see no more reason to do one thing than another. The solution is actually a path through a maze of decisions, and understanding how mice learn their mazes can help develop problem-solving strategies for students.

Modeling a Maze

Fig. 8-1 shows a particular maze, the Lashley III maze, that is frequently used in maze-learning studies on mice (Bovet, Bovet-Nitti, and Oliverio, 1969). On a trial, a mouse is put into the box labeled "start." After it passes out of the box, a gate closes on the box, so that the mouse can't return to its starting point. Thereafter, it wanders through the maze, in and out of the cul de sacs (I . . . VIII), until it reaches the goal and is rewarded with food. Then, after some interval, it is placed in the starting box for another trial. Each time the mouse enters a cul, it trips a switch that records the entry. In this way, the number of

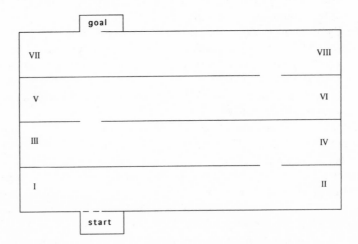

Figure 8-1 Lashley III maze. After the mouse leaves the starting box, a gate closes the box to prevent re-entry into it.

cul entries, or errors, on a trial can be automatically recorded. Most mice dramatically decrease their error rate after a few trials.

From our bird's eye view, the Lashley III maze doesn't look very mazelike, but from a mouse's perspective, it's far more complex. After passing out of the starting box, the mouse must turn left or right. If it turns left, it enters cul I and an error is recorded. If it turns right, it reaches the first opening without error. At the opening, the mouse has three choices: it can turn around, go into the opening, or go straight ahead into cul II. In my simple computer model of this situation, a mathematical, or cyber, mouse is given only two choices; it isn't allowed to turn around except after entering a cul. (Real mice often make a sensible fourth choice: they lie down and go to sleep, to the complete desperation of the graduate student whose thesis depends on her rodents' cooperation. Cyber mice never sleep.) If the cyber mouse goes into cul II, an error is scored and its direction is automatically reversed, bringing it to the opening from the opposite direction. At this point, it has the choice of going into the opening or going straight back into cul I.

The maze in Fig. 8-1 has twenty-one such decision points; these are shown as shaded triangles in in Fig. 8-2. Each triangle points in the direction the cyber mouse is moving at that point. For example, at point 4 a mouse has the option to go left or right after going into the first opening from the first corridor, whereas at point 5 a mouse has the option to go left or right after going into the first opening from the second corridor. It may seem unnecessarily complicated to treat each opening as six separate decision points, but, in fact, this is the simplest description of the maze-learning process and represents the lowest level of cognition. The cyber mouse is too dumb to understand that a left turn at 2 is the same as a right turn at 3. Or rather, I'm too dumb to write a computer program that models such high-level understanding.

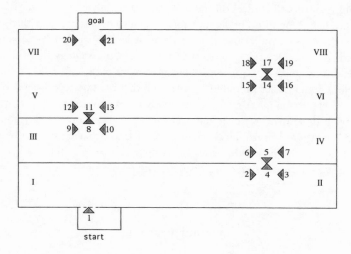

Figure 8-2 The twenty-one decision points of the model of the Lashley III maze in Fig. 8-1. Each triangle points in the direction that the mouse is moving at that point.

The alternative choices at each decision point lead to two other points, each with two alternative choices, and so on. The logic of the maze is represented by the rather awesome-looking decision map shown in Fig. 8-3. Each decision point in Fig. 8-2 is represented by a box in Fig. 8-3. The arrows leaving each box show the two choices at that point. For example, upon entering the first opening from corridor one (point 4), the mouse can turn left or right. A left turn takes it to point 10, where it has the option of going into the second opening or into cul III. A right turn takes it into cul IV, where it must turn around and arrive at point 7. Fig. 8-3 indicates this last choice by a heavy arrow from box 4 to box 7 with the cul number (IV) along side of it.

The computer model moves a cyber mouse from point to point according to the choice map in Fig. 8-3 and to probabilities assigned to the choices at each point. On its first trial, a cyber mouse has no knowledge of what choices are best, so all the probabilities are set to 0.5: equal probability of making either choice. Starting at 1, the computer generates a random number between 0 and 1. If the number is less than 0.5, the program branches to 2 (a right turn); if it's greater than 0.5, the program also branches to 2, but scores one error to represent a left turn into cul I. Say the random number was 0.36. Then the cyber mouse is at 2 without error. Another random number is generated, and the cyber mouse goes to 4 or 3 (via cul II) depending on whether the random number is less than or greater than 0.5. In this fashion the cyber mouse scurries around the maze. At some point, it may reach 10, go to 9 (via cul III), then to 6 and 5. Since the cyber mouse isn't even looking for a reward, it doesn't mind retracing its steps over and over again.

In this model, in which no learning takes place until after the first trial is finished, the average number of first-trial errors is sixteen. For real mice that were run through a real Lashley III maze, the average number of first-trial errors was fourteen. This is remarkable agreement, given the assumption the model makes that all choices on the first trial are equally likely. It indicates that real mice have little or no preference for going into openings rather than culs.[1] They behave to a large extent as a *tabla rasa,* at least as far as maze learning is concerned. The task of the model is to find a learning strategy that the cyber mouse can use to decrease its errors on subsequent trials.

Before doing this, it should be pointed out that both perfect recall and zero recall are bad strategies. A cyber mouse with a perfect recall would just repeat its initial bad performance over and over again. A cyber mouse with zero recall would keep the 0.5 probabilities for all choices, thus behaving randomly on all trials. It might, by chance, do better on some trials than on others, but, on the average, it would score sixteen errors. Learning requires some form of selective recall.

Operant Learning

Operant learning is learning that takes place in response to a positive reinforcer, or reward. It generally applies to situations in which a behavior pattern occurs

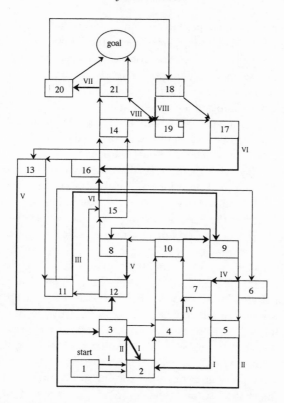

Figure 8-3 The decision map corresponding to the twenty-one choice points in the model of the Lashley III maze in Fig. 8-2. Each box represents a point in the maze where a decision between two alternative moves must be made. Each move is represented by an arrow pointing to the next decision point. The heavy lines represent decisions that bring the cyber mouse into and out of a cul and the Roman numerals correspond to the cul numbers in Fig. 8-2. For example, after going into the first opening, a cyber mouse has the option of going left or right. This is decision point 4 in the map. A decision to go right takes the cyber mouse into cul IV and back to the opening where it has the option (point 7) to reenter the first opening or head toward the second opening. A decision at point 4 to go left takes the cyber mouse directly to point 10. Each cul entry is scored as an error.

naturally with low frequency. With a proper reward schedule, the frequency of the pattern can be increased until the pattern becomes habitual.

Psychology textbooks tell the story of the psychology professor who had the habit of lecturing from the right side of the classroom, venturing to the left side for only a few minutes each period to write on the blackboard. One day his students decided to use operant learning to get him to spend more time on the left. In class, they looked down while he stood on the right, and looked up smiling when he moved to the left. As a consequence of this beaming reinforcer, the class got the professor to spend over half the period on the left side (Whaley and Malott, 1971).

Maze learning would appear to be quite different, because the subject has never exhibited the desired behavior; that is, the mouse has never taken the shortest path through the maze. Yet, from a mathematical point of view, the learning of a maze is amazingly simple. On its first meandering through a maze, no matter how convoluted the path, there are only two types of mistakes. The first is a simple entry into and out of a cul, and the second is a cycle that returns to the same decision point after a number of steps. On its first trial, a typical cyber mouse might make sixteen errors, three of which occur during a simple detour and thirteen of which occur during cycles.

By definition, if a cyber mouse visits the same point two or more times, it's in a cycle. It stays in this cycle as long as it makes the same choice each time it visits the point. It finally leaves the cycle by making the alternative choice. Thus, if a mouse visits the same point more than once, the last choice made at that point is the proper one. Optimal learning is achieved if, after completing the first trial, the probability for making the choice taken on the last visit to each point is changed from 0.5 to 1.0. For points visited only once, the last choice is the only one made, and for points visited more than once, the last choice eliminates cycles. On trial 2, there are only a few errors from single entries into culs, and all subsequent trials are run exactly the same as trial 2—there is no further learning. This optimal strategy can't eliminate individual entries into and out of culs; these remain as unexamined habits or "superstitions."

Critical to this optimal learning strategy is the forgetting of earlier choices made at the same point in favor of the last one made. Although we naturally tend to remember the last things we do, we can get confused. At least I can. On my fourth visit to a shop in a unfamiliar part of town, I came to an intersection where, on my third visit, I had made the correct turn. But prior to that, I had made the wrong turn, and now I couldn't remember which was which. Of course, had I made notes of my adventures, I wouldn't have had a problem. Note-taking is such a powerful memory expander that we forget how limited our unaided memory really is.

A more realistic model of how mice and people learn routes would allow for imperfect recall. After each trial, instead of setting some probabilities to 1 and others to 0, the probabilities are changed more moderately, in such a way that the number of errors gradually decreases with practice.

Goal-Gradient Hypothesis

The goal-gradient hypothesis of Hull (1943, 1952) changes the probability for repeating the choice made at a decision point by an amount that decreases with the distance (measured in steps) of the point from the goal. The computer runs a first trial with all the probabilities set to 0.5, deciding each step by the value of random numbers that the computer generates. After the trial, the probabilities of all choices are modified by amounts that decrease with the number of steps from the goal according to some reasonable schedule.[2] The last step, which goes into the goal, has the biggest change, say from 0.5–0.5 to 1–0; the next to last step, has the second biggest change, say from 0.5–0.5 to 0.75–0.25; the third to last step has the third biggest change, say from 0.5–0.5 to 0.67–0.33, and so on. Because of cycling, the same point may have been visited several times during the first trial, but the last visit is closest to the goal, so the last choice made at the point is reinforced more than earlier choices.

After these modifications, the cyber mouse is run through the maze again. Generally, it does much better. Fig. 8-4 shows the learning curves for cyber mice with two different reinforcement schedules, labeled Cyber 1 (black circles) and Cyber 2 (black squares).[3] Each curve is the average of 100 cyber mice. Note that on trial 1—where all the probabilities are 0.5 for both schedules—the average number of errors given by the two schedules are fifteen and seventeen. On the first trial, there is no learning, so this difference is purely the result of the statistical fluctuations that remain even when averaging the results of 100 cyber mice. After seven trails, both schedules significantly reduce the errors, but Cyber 1 clearly does better than Cyber 2.

Fig. 8-4 also shows the learning curves for two strains of real mice, labelled BALB/c (white circles) and CBA (white squares). These data were recorded in 1969 in a study designed to test whether genetics affects learning ability in mice (Bovet, Bovet-Nitti, and Oliverio, 1969). The mice in a given strain are the product of hundreds of generations of inbreeding and are genetically identical. Any difference in performance among individuals of the same strain should be due to statistical fluctuation, just as is the difference in performance

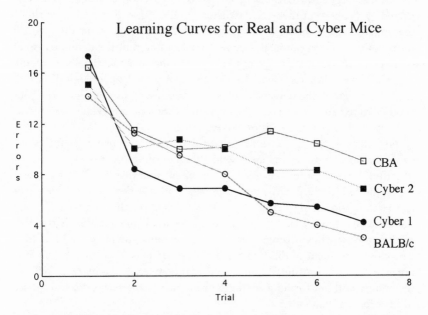

Figure 8-4 Learning curves for two reinforcement schedules of cyber mice (black circles and squares) and two strains of real mice (white circles and squares). The curves of the real mice are each the averages of 16 individuals, whereas the cyber-mice curves are each the average of 100 cyber mice.

of the cyber mice with the same reinforcement schedule. Any difference between the average performances of the different strains should be due to the genetic difference between the strains, just as the difference between the average performance of the two sets of cyber mice is due to the difference in their reinforcement schedules.

Fig. 8-4 shows that the model produces learning curves of the same general character as those produced by real mice. There's more statistical fluctuation in the data for the real mice than for the cyber mice because the real data are the averages of only sixteen individuals in each strain, whereas the cyber data are the averages of 100 cyber mice for each schedule. Repeating the experiments with sixteen different real mice of the same strain would yield learning curves that, point by point, differed from the ones shown by one error, more or less. This means that the difference between the learning curves for strain CBA and model 1 on the one hand, and strain BALB/c and model 2 on the other, are not statistically significant, though the differences between the two strains and two models are. The model, therefore, does a creditable job in explaining how mice learn their way through a maze.

Human Learning

As we have argued in Chapter 2, science is a collection of related data connected by a consistent theoretical structure. In this case, we are calling our theoretical structure a model because it doesn't say anything about the physical nature of the learning process. The probabilities stored in the computer don't correspond to any similar quantities stored somewhere in the brain of the mouse. Rather, they represent the net result of several physiologically based tendencies, such as the tendencies of a mouse to distinguish one location from another, to remember what it did at a given location on a previous trial, and to make its decision on the basis of past experience.

Nevertheless, the model is essential for drawing any general conclusions about learning, and the data are essential for validating the model. For example, from the data on real mice it may not be obvious that the first-trial scores are all the result of a purely random process, and that there is no evidence of any learning occurring before the goal is reached on the first trial. But the model, which doesn't change any probabilities until after the goal is reached, predicts precisely how many errors will be made on a purely random first trail. The close agreement of this number with the real-mice data supports the no-learning conclusion. Similarly, since a cyber mouse learns the maze as well as a real mouse without forming anything like a map of the maze, the real mouse needn't form a map either.

Applied to human beings, the no-learning conclusion contradicts the constructivist doctrine that process is more important than outcome. For a constructivist, a child learns through the process of exploration, whether or not the exploration results in any objective success. The model requires success to promote learning, and the no-map conclusion means that complex tasks can be learned without understanding. This stretch from mice to men isn't as preposterous as it seems. Both are foraging creatures, who make their living by exploring their environment for food. The best strategy for this is to move about randomly until food is found, and then to return to the fruitful location as long as it has food. An understanding of the seasonal variations of fruiting trees isn't necessary for this, though with such understanding human beings advanced beyond the limits of operant learning.[4]

As we showed in Chapter 2, even the simplest science activities have more erroneous pathways and cul de sacs than teachers, untrained in science, realize. In mucking about randomly, a student learns as little as a mouse does while meandering about the maze on its first trial. Only when the student reaches a goal, such as getting an experiment to agree with an equation, does the whole enterprise begin to make any sense. The path to this goal can be extremely tortuous, just as is the mouse's first run through a maze. The teacher's role is to help all the students in a class reach the goal within reasonable time limits. An experienced science teacher knows that some detours are so wasteful of time and energy that students should be warned against them, whereas other byways might be left for the students to explore.

During a semester of physics, a student might do twelve or so laboratory experiments, just like a mouse that is required to run the Lashley III maze seven times. The critical difference is that the mouse runs the same maze each time—how else could it learn it?—whereas the student does a different experiment each week. We would expect, on the basis of our analogy, that students learn only what is common to all of the labs—data tabulation, graphing, analysis—and not the physical principles that vary from experiment to experiment. In my own laboratory (Cromer, 1994), I try to keep the principles to as few as I can, and to repeat them as often as I can. It may seem far-fetched to compare a student doing a physics experiment with a mouse running through a maze, but only to someone who has never taken a physics laboratory course.

There is one important difference: The goal of most physics experiments isn't as evident to a student as a food pellet is to a mouse. If the prediction for the period of a three-meter-long pendulum is 3.5 seconds and the measurement is 3.1 seconds, has the goal been reached? The teacher might engage a class in a discussion of this question, but this would require a fairly sophisticated understanding of precision and experimental error. Most students, even those with considerable experience with experimental methods, are inclined to dismiss a 0.4-second differences between prediction and experiment as "experimental error," and go on to the next experiment. But if this is allowed to pass, the students will have learned a procedure with a critical error in it. Worse, they will get the false impression that theory and experiment never agree very well. The graduate students who teach our undergraduate physics laboratories sometimes complain to me about experiments giving poor results, when, in fact, routine procedural errors have gone uncorrected. These young instructors feel they have reached their goal when the laboratory period is over, not when they have helped their students get it right.

After a mouse has reduced its errors on a maze from sixteen to four, it is still making four ritualisticlike turns into culs before proceeding to the goal. This is like the behavior of Konrad Lonrez's greyleg goose who, upon entering his house, always ran past the stairs to a window before returning back to the stairs and going up to its room. This is the route it took the first time it came into the house, and it kept to it thereafter. Once, when it forgot the ritual, it got very distressed five steps up the stairs, and came back down, went to the window, and then went happily up the stairs (Lorenz, 1966).

The goal-gradient-operant learning model eliminates cycles—sequences of steps that come back to the same point—but it can't eliminate all of the individual entries into and out of culs. The unnecessary steps that a trained mouse makes on its way to the goal are of little consequence to the mouse. Indeed, routinely checking out unproductive byways has good survival value, because, in the real world, the sources of food are ever-shifting. For students in a physics laboratory, some errors are like turns into cul de sacs in that they don't noticeably affect the outcome of the experiments. Being unnoticed, the students can't, on their own, correct them.

Critical Thinking

The term "critical thinking" is commonly used in education to refer to some sort of high-level cognitive process. For some, it's the process involved in solving unusual word problems: "If a string wraps tightly around the earth at the equator, how much longer must it be to wrap around the equator while staying exactly one foot above the earth?" For others, it's the process of developing arguments to defend a thesis. In our adversarial society, there is much reward for being able to argue either side of a case.

Common to both views is the notion that critical thinking is thinking about thinking, or, to use more operational language, that it's a procedure of procedures. Our cyber mice have a procedure for finding a short route through a maze. A critical-thinking mouse would have some higher-level procedure to test whether its short-route finding procedure had found the shortest route. As far as we know, mice don't modify their behavior this way, but people sometimes do.

The expanded capacity of working memory undoubtedly plays a critical role in this. Human beings are believed to be able to work with four to six "chunks" of memory at a time, compared to perhaps two for animals (Waldrop, 1987; Cromer, 1993). This seemingly insignificant increase is enough to allow an automobile mechanic to find the tool to operate on a particular part, because he can remember the part and the operation while searching for the tool.

Language further enhances the functionality of working memory by supplying words for ever more complex ideas and relationships. Only in language can we express such general relations as "The circumference of a circle is pi times its diameter." Even this may be too large for most people's working memory. My working memory can hold only the formula "$C = \pi D$." But it's precisely through such compressions and abbreviations that human beings are able to do so much with so little.

Writing is another major aid to thinking, since it allows us not only to save the contents of many working memories, but to reduce these contents to single chunks, allowing several contents to be thought about at once. I can stare at $C = \pi D$, while thinking about the string encircling the earth. The problem doesn't give either C or D, which is what makes it so challenging. But neither does it ask for the length of the new string; it only asks how much longer it is than the old string. Very few students get this far, but if they do, they might draw a picture like that in Fig. 8-5 and realize that if D is the diameter of the earth, then the diameter of a string that encircles the equator one foot above the earth is $D + 2$ feet. With the formula $C = \pi D$ in front of them, they may see that they can write the enlarged circumference C' as

$$C' = \pi(D + 2) = \pi D + 2\pi.$$

From here, it's a few mathematical steps to the goal:

$$C' - \pi D = C' - C = 2\pi \text{ feet.}$$

Surprisingly, a string that encircles the earth one foot above the equator is just $2\pi = 6.28$ feet longer than a string that is wrapped tightly about the earth at the equator. The answer doesn't depend on the size of the earth.

Solving such a multiple-step problem is much like running through a maze. The sequence of steps from the given to the goal isn't clear and there are many cycles and cul de sacs that one can get into. Although there's a logical justification for each correct step, there are many logically correct steps that don't lead to the solution. Not until the problem is solved can one be certain that one has found the correct sequence of steps.

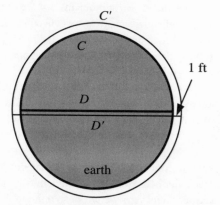

Figure 8-5 One string of length C is wrapped tightly about the earth at the equator, while another string of length C' encircles the earth one foot above the equator.

Mazes and problems also share the obvious characteristic that there are fewer decisions, and hence fewer potential errors, the closer one is to the goal. This suggests that most problems may be easier to solve if one proceeds backwards from the goal to the given. This requires conditional thinking: if I knew the answer, then I could calculate the step just before the answer; and if I knew the step just before the answer, I could calculate the step just before that, and so on back to the givens in the problem. For example, the goal in the string problem is the difference in the lengths of two strings, or the difference $C' - C$ in the circumferences of two circles. Since the circumference of a circle is pi times its diameter, the next step back from the goal is to write

$$C' - C = \pi D' - \pi D$$

where D and D' are the respective diameters. Knowledge of algebra might suggest factoring out the common factor of p, and writing

$$C' - C = \pi(D' - D).$$

Since the given is the height of the longer string about the earth, the last step suggests Fig. 8-5 and thus that $D' - D = 2$ feet, which completes the route from goal to given.

I call critical thinking a procedure of procedures because each step in a critical-thinking problem is itself a learned procedure, whether it's factoring π or replacing the symbol C with the symbol πD. Without a firm knowledge of these simpler procedures, it's not possible to learn more complex procedures that involve them. I'm belaboring the incredibly obvious point that critical thinking requires extensive prior knowledge because constructivist edubabble teaches just the opposite.

> Problem solving as discussed here [Review draft of the Massachusetts Curriculum Frameworks for Mathematics] means something quite different from typical word problems. . . . Most problems that are encountered in real life are quite complex—there are many factors affecting the situation, there are many ways to view and approach the problem, there are many strategies for solving the problem, there is frequently more than one solution, and several criteria for what is the "best" solution.
>
> [For example:] How do I choose the best car for my family? Should I buy now or wait another year and risk substantial repair bills? What are the most important factors for me to consider in selecting a car? What weight should I give to each factor? (Massachusetts, 1994)

Although this may seem reasonable, it's actually suggesting that mathematics teachers replace well-defined exercises in problem solving with bogus exercises in decision making. There's no way to determine logically or empirically what "weight" to give to such competing factors as the pleasure of owning an expensive car and the pleasure of having extra cash in the bank. All that students can do is guess at their personal "weights," undermining the growth of objectivity and logical thinking that the study of mathematics is supposed to promote.[5]

Teachers across the country are being told that problems with clear outcomes must be replaced with "open-ended problems" that have no "best" solution. The idea behind this is that students will learn critical thinking by running around in indefinitely large mazes with ill-defined goals. From the very process of running around in circles, they say, students will learn to think critically. Yet there isn't any evidence that learning ever takes place without the reward of tangible success, or that complex procedures can be mastered before mastery of the simpler procedures of which they are composed.[6]

Educators champion goalless maze running because it doesn't require teachers to have subject-matter competency. Anyone can propose ill-defined problems that have "several criteria for what is the 'best' solution." Many of the educational authorities leading the constructivist movement in science education have no knowledge of science at even a middle-school level. Nor do they feel that they should have, since education is no longer to be about learning anything. It's to be about the content-free, criteria-free solution of poorly defined problems.

As crazy as this sounds, it is consistent with constructivism's denial of objective truth. Constructivism forbids teachers to give focused lessons, because this would impose the teacher's view of truth on students. Since there is no truth, the teacher should have no privileged position in the classroom.

Guided Inquiry

But without knowledgeable guidance from their teacher, students are truly like mice in a maze. Each will arrive at his own version of the goal with his own set of errors and misconceptions. Nothing real is achieved, so nothing real is learned.

Consider the case of the students who found the period of a three-meter-long pendulum to be 3.1 seconds, when their prediction, based on analysis of measurements they made with shorter pendulums, said it should be 3.5 seconds. The whole point of the lesson is to show how, with some simple analysis, one can connect the results of different experiences. This is the very essence of science, as we have discussed it in Chapter 2. Without better agreement between prediction and experiment, the point is lost.

It's imperative that the discrepancy be resolved within the time allotted to this topic. This means the teacher must thoroughly understand the correct procedure, and the most frequently made errors. It isn't necessary to announce the error, but neither is it reasonable to expect most of the students to find it by themselves.

One might ask a student to repeat her measurement in front of the class. Her procedure is to measure the time it takes for the pendulum to complete ten oscillations, starting and ending when it reaches its farthest left-hand position. As the pendulum reaches its left-hand position for the first time, she starts her stopwatch and her counting: 1, 2, 3, . . . up to 10. She finds 31.1 seconds, or 3.11 seconds for one oscillation. The class is asked to spot her error. So is the reader.

Most students say she started or ended her timing a little early or a little late. This could cause a few tenths of a second error in the total time for ten oscillations, but since the total time is divided by ten, this error contributes only a few hundredths of a second to the error for the time for one oscillation. We are looking for a much bigger error.

Finally, one student gets it. There are only nine oscillations between one and ten. By starting the count at one, rather than zero, the student actually measured the time for nine oscillations. Dividing 31.1 seconds by 9 yields 3.45 seconds for the time of one oscillation, in good enough agreement with the prediction of 3.5 seconds.

This is just one example of the many small, but critical, errors that are possible in even a simple experiment. The teacher's role is to help his students through an experiment, assisting them when they get stuck, and preventing them from making ruinous errors. This requires considerable training and experience. Even simple experiments are "real-life" situations with real complexity; more so than is an ambiguous exercise on choosing the "best" family car. The teacher shouldn't be just a nudging facilitator of his students' self-learning, but must be an expert who can successfully guide them through a maze of confusion and error.

Social Learning

After seven runs through the Lashley III maze, the BALB/c mice in Fig. 8-4 still made an average of four errors. The cyber mice behave similarly, and a detailed examination of the computer-generated data shows that there is wide variation among individuals. After seven, or even twenty trials, some cyber mice make one or no error and others make seven or eight errors. Undoubtedly, real mice show as much variation. Even mice who make the same number of errors, make them by going into different culs in different orders, so that out of 100 cyber mice, few learn to run the maze along the exact same route. Each mouse, in effect, has constructed it own understanding of the maze.

Such personal, or idiosyncratic, understanding is sufficient to the needs of many creatures in many situations, but it's far from objective knowledge. To go beyond personal understanding, the mice would have to talk to one another about their experiences. They would argue about how many cul de sacs there were to the goal, some saying eight, some saying four, and few saying one or zero. These differences could divide the mice into sects, each preaching a different view of the maze. Or, if mice preferred to avoid cul de sacs, they might all learn the route that reaches the goal without encountering any cul de sacs. In this case, communication among individuals has led to a group decision as to which is the best route.

Science is the ultimate refinement of this communication process, with open discussion of competing routes being socially mandated and division into sects being discouraged. But scientific understanding of the maze occurs only when the Copernicus of mice comes along with a theory that explains all the individual routes as possible variations within the same objective reality. Almost all scientifically inclined mice accept the theory because it enables them to navigate freely around the maze, while a few continue to seek alternative routes and explanations. These renegades are tolerated, just in case there is more to the maze than is explained by the accepted theory.

Of course, education isn't research, and most of what students are expected to learn is established knowledge. Indeed, there is little point in taxpayers supporting the learning of unestablished knowledge. Nevertheless, what students learn is new to them, and so learning involves a process of discovery. This is clearly seen in some eighth-grade classes in Boston that are using graphics calculators to study algebra. The calculators are like a maze: they have a dozen buttons that lead to different menus, and each menu has many choices, including going to still more menus. Although the buttons and menus are labelled, these mean nothing to beginners. Guided by their teacher, the students try to plot lines with different slopes, but soon many are lost among the menus. They try this, they try that. One student discovers that the menu produced by the "Y =" key depends on what has been selected on the "MODE" key. This news is triumphantly passed on on to others. Through a complex process of instruction, exploration, and intraclass communication, the students rapidly gain familiarity with the calculator.

Many children seem to lose their exploratory instincts as they get older. Most educators attribute this to the deadening influence of schooling itself, but I tend to see it as a part of the natural socializing process. For a social animal, conformity, rather than originality, is the key to survival. Children rapidly commit themselves to the language and rules of their group, because their survival depends upon it. From birth through high school, a student spends less than eight percent of her life in school; most learning, for better or worse, takes place outside of school hours.

Students do show interest and exploratory behavior in situations, such as the calculator class or the science laboratory, where they encounter new situations about which they haven't committed themselves. It's also been found that most students are more creative when they work in small groups than when they work alone. Students have always worked in pairs in the laboratory because of the lack of equipment, but teachers are now deliberately forming students into learning groups, in and out of the laboratory, because group work is more imaginative and inventive than most individual work. In the physics laboratory at Northeastern, students seem to do better when working in groups of three rather than two.

I became convinced of the value of having students work in groups after I organized the students in my Introduction to Science course for nonscience majors into groups of three, and required each group to submit a single paper for each assignment. Previously, when students had been assigned individual work that involved critical thinking—Measure the thickness of a page in your textbook.—few bothered to do it. With group assignments, which forced students to discuss the problems among themselves, there was 50 percent compliance.

The problem of measuring the period of a three-meter-long pendulum turned out to be more challenging than I had expected, because most ceilings are about 2.5 meters (8 feet) above the floor. Some groups suspended it out a window, others down a stair well, and still others settled for a 2.5-meter-long pendulum. There's no question that this assignment, which was rather fun to do in small groups, wouldn't have been attempted by individuals.[7]

As a result of this experience, I increased the social contact among students in the physics laboratory by having them work in groups of three, rather than the traditional pairs. This was seen as regressive by some faculty, who have a romanticized image of the solitary scientist discovering the secrets of the universe alone at night in his study. But scientists share ideas all the time, and students also do much better if they are involved with a learning group. A group's solutions to problems are generally better than the solutions of the best problem-solver in the group (Heller, Keith, and Anderson, 1992). Working in groups enhances objective thinking, since to prevail, an idea must make sense to everyone in the group. Indeed, some sense of objectivity is required before group work is possible. Young children, and some adults, are too egocentric to cooperate on projects with peers. They simply don't accept the validity of other people's opinions. Learning the social and cognitive skills needed for cooperation and team work is seen as one of the positive benefits of having younger students work in groups. Fortunately, by the time most students arrive in college, they do have these skills.[8]

But a group of three is still just a group of three, and it can come up with procedures and results that are wrong, that is, that are contrary to the procedures and results established over many years by millions of scientists, teachers, and students. A knowledgeable teacher is still required to steer the work in the right direction. The reward of a physics experiment is a result that agrees, within acceptable limits, with the prediction of a theory. Failure to achieve this reward compromises the value of the whole learning experience.

Fixation

The cyber mouse learns a maze by increasing its probability of repeating its most recent behavior. Since these are the decisions that turned it into the goal or got it out of a cyclical sequence of moves, this process quickly adapts the mouse to the maze. Of course, in real life people often fixate on an early experience, and fail to learn from later ones. It's easy to program the cyber mice so that their probabilities of repeating earlier experience are increased more than their probabilities of repeating later ones. Now, after its first run through the maze, a cyber mouse's probabilities for getting into cycles are greater than its probabilities for getting out of them. On its second trial, it stays in cycles longer than on the first trial, and its probabilities for doing so are increased further. This process continues until the probabilities of getting out of cycles are nearly zero; the cyber mouse is now trapped in the maze, repeating the same cycle of moves over and over.

The analogy to compulsive-obsessive behavior in humans is interesting. Indeed, much of psychotherapy is devoted to helping people to get out of repetitive patterns of behavior. Similarly, the resistance to learning displayed by many poor learners may be an early egocentric deviance that becomes fixed. These students becomes trapped in cycles of nonadaptive and maladaptive behavior.

The schools are no more to be blamed for this than are families and friends. What is clear, if this analysis has any merit, is that students may be able to free themselves from their own entrapping behaviors if they are put in situations with unfamiliar decision points. This could be smaller classes with personal attention and concrete learning experiences. It's remarkable just how abstract, fragmented, and uninformative many textbooks are. For example, a lesson on battery and bulbs may implicitly assume that all students know how a light bulb works. Yet the whole point of the lessons is missed if a student doesn't understand that there is a single wired path from a metal nipple at the bottom of the bulb, up to the filament, and down to the metal threads near the base.

I have asked students in a seventh grade class and in an Adult Basic Education class to draw what they think is inside a light bulb. In both classes, there were a few who had some concept of a filament and many who just drew a single wire with sparks jumping off of it. None could draw how the wires connect to the base. I then passed around a light bulb from which I had broken off the glass. The simple act of examining the inside of a light bulb is for many a new type of school experience. It may be the first time they have ever looked inside of anything, or realized that what goes on in school has any connection to what goes on in the real world. Such unfamiliar experiences, when matched to the students' current cognitive level, can break negative expectations and nonproductive cycles, allowing students to move on to higher levels.

But all students, even in the same grade, aren't at the same level, and they don't all progress at the same rate. What is the source of this variation? Why do some students succeed against all odds, while others lapse into despair and degradation in the midst of hope and opportunity? Is it the Grace of God, as Calvinists believe, or is it the dance of genes? Some sociologists tell us it's all a matter of teachers' expectations: if only teachers expected everyone to be above average, they all would be.

The next chapter tackles the thorny issue of human differences. How immutable are genetic differences? Is intelligence a meaningful scientific concept? And if so, how should schools handle differences in intelligence? These questions can't be avoided, because they stand at the nexus of science, philosophy, and education. It's important to develop a language in which serious discussion and debate can be conducted without the thoughtless anger and vitriolic rhetoric that the subject seems to engender. This is the "Mission: Impossible" of the next chapter.

9

Human Variation

All human beings have a common ancestry, yet each is different from the other. This natural variation among individuals, which exists in all sexual species, was recognized by Darwin as the raw material for evolution. Yet as natural as human diversity is, discussion of it is fraught with angst and confusion. Modern democracy rests on the concept of a social contract among equals, and variation implies inequality. To many, any discussion of variation is seen as a threat to democracy itself.

But the opposite case can be made just as easily. If we were all equal, we wouldn't need democracy, because any one of us could speak for all the rest. It's because we are different that we are each entitled to an equal voice in public affairs. And because democracies can tolerate variety better than other forms of government, they have a greater capacity to evolve to meet changing conditions. Societies, as species, can't predict what the next catastrophe will be, so they had best keep all their options open if they wish to survive in the long run.

Origin of Variation

In each cell of the human body, there are approximately 100,000 genes located on forty-six chromosomes, twenty-three coming from the mother and twenty-three from the father. Each gene is a segment of a DNA molecule that codes for the manufacture of a specific protein. All genetic variation, as far as we know, arises from variations in the coding sequences and, consequently, in the resulting proteins. It has been estimated that perhaps 1,000 human genes show such variation, the remaining 99,000 or so being functionally identical for all human beings on earth.

Twenty-two of the maternal and paternal chromosomes are matched (homologous) pairs, containing the same, or nearly the same, genes in the same positions along their nearly identical DNA. Replication errors, or mutations, create slight differences between some of the corresponding paternal and maternal genes. The twenty-third paternal chromosome may be an X or Y. If it's an X chromosome, it will match the twenty-third maternal chromosome, which is always an X. The resulting XX individual is female. The Y chromosome contains the maleness gene (or genes), and an XY individual is a male. Thus females have two copies of all genes, whereas males have only single copies of the genes on their X and Y chromosomes.

Sex is a cyclical process in which chromosomes from two mature individuals are combined to create a new individual. When this individual matures, the paternal and maternal chromosomes it inherited are segregated into its egg or sperm cells. The segregation process is random, each sex cell receiving a different combination of paternal and maternal chromosomes. When this individual mates, the segregated chromosomes in one of its sex cells recombines with the segregated chromosomes in a sex cell of another individual, to create a third unique individual. This cycle of recombining and segregating chromosomes is much like shuffling and dealing cards; both processes create endless numbers of combinations of the same units. From the point of view of egg and sperm cells, the loves and lives of individuals are just the mechanism they use for mixing up their chromosomes. Or, as the geneticists like to put it, a chicken is an egg's way of making another egg.

By itself, segregating and recombining of chromosomes only mixes maternal and paternal genes that are on different chromosomes. But homologous chromosome pairs often exchange chromosome fragments, so that some genes that were on a maternal chromosome are switched with genes on a paternal chromosome. Thus, over time, all the genes in human beings can be mixed in any combination, though genes close together on the same chromosome segregate less frequently than do genes on different chromosomes. If there were only two forms (alleles) of 1,000 genes, there would be 2^{1000} distinct combinations. This is far larger than the total number of sperm and egg cells ever produced by all the creatures that have ever lived, or ever will live.[1]

Sex is primarily about variation, not reproduction. There are asexual species of plants, insects, fish, and lizards that are exclusively female (Crews, 1994). In these species, the eggs have the full double-dose of chromosomes and develop into daughters that are genetically identical to their mother. Whatever variation there was among the founding mothers of the species is passed on to their daughters, creating clones of identical individuals. That is, all the descendants of one founding mother are identical to one another, though they may show some variation from the identical descendants of another founding mother. If a mutation occurs in an individual gene of one individual, this will be passed on to all of her daughters, creating a new clone (Maynard Smith, 1978).

Not only is the asexual reproduction of females biologically possible, it is, from a sociobiological point of view, highly desirable because it eliminates the risk and expense of finding a mate and then having to compete with him for food. Yet, in spite of its competitive cost, most species segregate and recombine genes at some point in their life cycle—even some species of bacteria do it.[2] This must be because the cost of sex is outweighed by the advantage sex conveys by being able to mix mutations that arise in different individuals. For example, ten mutations of ten genes in an asexually reproducing species create ten clones, but segregation and recombination has the potential of creating $2^{10} = 1,024$ distinct combinations of mutated and unmutated genes.

The consequence of this genetic diversity is shown by the ability of some bacteria to develop resistance to many, if not all, antibiotics. Bacteria with a certain mutant gene A' may not compete effectively with bacteria with "normal" gene A, so most bacteria have gene A. But if gene A' produces immunity to penicillin, then, under chemical attack, A' bacteria thrive. Before penicillin, A bacteria thought themselves far superior to the rare mutations A'. But under new circumstances, the last shall be first: A' becomes "normal" and A becomes the rare mutant weakling. Likewise, if B' protects against tetracycline, then B' bacteria thrive in a tetracycline world. Sexual bacteria occasionally recombine and segregate their genes, so that an A' and an B' bacterium could produce a A'-B' bacterium that is resistant to both penicillin and tetracycline.

It's important to realize that a particular mutation that is deleterious under "normal" circumstances may be advantageous under abnormal circumstances. This is illustrated in human beings by the well-studied case of the sickle-cell gene, a mutation that causes abnormal red blood cells. Before modern medicine, individuals with two sickle-cell genes—one from each parent—died in infancy. Individuals with one sickle-cell gene and one normal gene have less abnormality and can live, but are not as healthy as individuals with two normal genes. In most environments, sickle-cell individuals don't thrive in competition with normal-cell individuals, and the sickle-cell gene is rare. However, individuals with one sickle-cell gene are more resistant to malaria than are individuals with normal cells, so in areas of Africa where malaria is endemic, this abnormality becomes an asset and the gene is common. If 20 percent of the genes in a population are sickle-cell, four percent of the population will inherit two sickle-cell genes and die, while 32 percent will have one sickle-cell gene and be, on the average, healthier than the malaria-infected "normals." This classic example from population genetics shows that human diversity protects the group, not the individual (Dobzhansky, 1973).

Moreover, the human species, with its vast genetic diversity, is more thoroughly armed against annihilation than is any single group within the species. Disease has destroyed whole tribes, but the species has survived. As deadly as smallpox was, there were individuals who resisted it. As deadly as AIDS is, there appear to be resistant individuals. From these, science may learn how to provide resistance to everyone.

Nothing in this discussion requires that there be just two genders, or that males and females be genetically different, that is, XX females and XY males. Indeed, in many species this is not the case, the sex of the individual being determined by temperature, or other environmental factors during embryonic development (Crews, 1994). The whys and wherefores of the many ways in which organisms segregate and recombine their genes is a fascinating subject for research and speculation (Margulis and Sagan, 1986; Halvorson and Monroy, 1985; Maynard Smith, 1978; *Science,* 1995), but well beyond our current concern.

Race

In the loyalty–rebellion model of Chapter 5, it is argued that human group size is limited by the tendency of rebellious individuals to break away and form a new group when the size of their parent group gets too large. Over many generations, group formation spreads people geographically, and hostility among groups prevents individuals from traveling very far from their home group. Over time, people only twenty kilometers apart can become culturally and sexually isolated from one another. Their customs and languages drift apart, and they eventually become genetically distinguishable. At this point, the people can be said to belong to different races.

The term "race" is most meaningful when restricted to small endogamous groups, that is, to groups whose members mate with one another and don't mate with members of other groups. The barrier to outside mating may be physical isolation, or it may be cultural or racial prohibitions. These barriers are never absolute, so the concept of race is as fuzzy and inconsistent as people are. But natural barriers do exist, and artificial ones are constantly being created, so over time, the continual splitting-off of groups creates ever more races, resulting in a world filled with ethnically and racially diverse tribes. The longer the process goes on, the greater the diversity. One of the principal reasons for believing that *Homo sapiens* originated in Africa is that the peoples of sub-Saharan Africa are the most genetically diverse on earth.

When some tribes migrated out of Africa 100,000 years ago, they carried with them only a fraction of this diversity (Fischman, 1996). Over time, these emigrants also became more diverse, but the process was also continuing back home as well. Today, by studying the variations in the DNA of peoples around the world, geneticists hope to be able to trace the footsteps of our ancestors.[3]

Between 1500 and 1850, the slave trade transported tens of millions of Africans to America—a second African diaspora. Although the slaves were taken from many tribes and races, they didn't represent the full diversity of Africa. In the United States, African-Americans—the descendants of African slaves—share a common gene pool that is the blending of the gene pools of their ancestral races, both African and European (Eysenck, 1971).

European-Americans have been mixing it up genetically for about as long as African-Americans, but nineteenth-century European immigration brought in many different endogamous groups which partially maintain their identity and endogamous preferences. Thus the European-American population still consists of a number of races with small genetic differences. These manifest themselves mostly in differences in the prevalence of certain genetic diseases. For instance, Jews of eastern and central European origin have a greater than average incidence of Tay-Sachs disease and one percent of this population carry a particular mutation on a particular gene linked to breast cancer (Kolata, 1995; Struewing, Abeliovich, and Brody, 1995).[4]

Racial Differences

There are two common misconceptions about racial differences that are important to discuss. The first is that there are no significant differences among the races, and the second is that if there were such a difference, it would necessarily be immutable. Politically, it is felt that any important racial difference, like poor school performance, must have a correctable social cause, rather than an incorrectable genetic cause. Yet nothing is more obvious than that there are important racial differences and that the problems they create are correctable.

Low-pigmented skin evolved as human beings moved from the tropics to the northern latitudes. There, where the rays of the sun always strike at an oblique angle, the lack of pigmentation is of little concern, whereas even a few hours in the tropical sun would severely burn the unprotected body of a Scandinavian. As it is, a few hours of New England noonday sun on the Fourth of July often burns a northern European's body. Sunburn and sun poisoning are a joint consequence of the European custom of sunbathing and the racial lack of sufficient skin pigmentation to protect against the sun. Fortunately, science and sunscreen save the white man's hide.

The problem is more serious for farmers and construction workers who work outdoors in the summer, because chronic exposure to sun causes skin cancer—the most common cancer among whites. With the weakening of the ozone layer and the increasing amounts of ultraviolet radiation reaching the earth, the health problems associated with low-pigmented skin will become more prevalent. The situation isn't hopeless, but corrective action and changes in habits will be necessary.

Highly pigmented skin has its own problems in northern latitudes. Vitamin D, which is necessary for normal bone growth, is found in some foods (fish liver oil and egg yolks), but is mostly produced in skin exposed to sunlight. White skin is highly efficient in producing vitamin D this way, so that even the cheeks of a Scandinavian child exposed to the weak winter sun will produce sufficient amounts of D for healthy bones. Highly pigmented skin is less efficient, not having evolved in a low-sunlight environment. Consequently African-American children raised in the North are subject to rickets, a deformity of the bones caused by malnutrition and vitamin D deficiency. Sixty years ago rickets was a major health problem of black children in the United States, which has been largely corrected by better nutrition and the addition of vitamin D to milk.

Genetic problems such as these are inevitable because of the movement of people from their traditional environments to new circumstances. As other problems are identified, appropriate corrective measures can be taken. At our present state of knowledge, we can change people's chemistry easier than their behavior.

Testing Differences

Nowhere are human differences more apparent than in the classroom, where students are constantly being tested on what they have learned. The purpose of these tests is to goad students into studying and to provide an accounting of classroom progress. The question that nags every educator is "What do tests measure, if anything?" One concern is that a student's score on a test may measure nothing more than the score itself. To be meaningful, the score should at least predict how well the student would do on another test on the same subject. Another concern is that a student who has practical or conceptual knowledge of a subject may do poorly on a test of facts, whereas a student who knows the facts may understand very little. But how can one test understanding except with another questionable test?

In the days when California still had money, research was undertaken to design alternatives to standard multiple-choice tests that could be administered statewide and that would be in accordance with the constructivist notion of learning and teaching. These tests, or assessments as educators prefer to call them, were "concrete, meaningful tasks . . . scored so as to capture not just the 'right answer,' but also the reasonableness of the procedure used to carry out the task or solve the problem" (Shavelson, Baxter, and Pine, 1992). One of the tasks asked of fifth and sixth graders was to determine what was inside six sealed boxes by connecting a battery and/or a small light bulb to terminals on the outside of the boxes. Inside a box, the terminals were connected by a plain wire, a battery, a bulb, a battery and a bulb, two batteries, or nothing. As a child took the test, she was observed by two researchers who scored her independently. There was high correspondence between the scores of the observers, indicating that objective scoring criteria had been established.

Such an assessment, in which a student is carefully observed while she undertakes a meaningful investigation, was considered to be the benchmark against which more affordable tests were to be compared. For possible use on a statewide basis, a computer version of the sealed-box test was developed in which pictures of batteries, bulbs, and mystery boxes could be moved on the screen with a mouse. Whenever a circuit was connected, the bulb or bulbs glowed bright or dim, or didn't glow at all, as would happen in real life. The computer also scored performance as the live observers had. To compare this test to the live test, the same students had to take them both.

A group of seventy-three students who had had hands-on experience with battery and bulbs were given both the live and computer test. The average on both tests was 3.7 (out of six), and the correlation between the scores was a very respectable 0.9. A third of the students achieved mastery of the subject, correctly identifying five or six of the six boxes on both tests, and only 15 percent identified three or fewer on both tests. However, half of the class achieved inconsistent results on the two tests: some students identified more boxes on the live test than the computer test, and some just the reverse. In fact, half the class would have been judged to have achieved mastery based on the scores from either test alone. For example, among the twenty students who scored six on the computer test there were five who scored three on the live test, and conversely, among the twenty-seven students who scored six on the live test there were seven who scored three or less on the computer test.

Based on their results, Shavelson, Baxter, and Pine concluded that the knowledge and technology for a nationwide system of performance testing doesn't exist at the present time and that there's no empirical evidence that performance testing will improve education even if it could be implemented (1992). They warn of the difficulty of developing and implementing the system of voluntary national tests that President Bush advocated in the America 2000 program (Bush, 1991).

Individual and Group Differences

Tests are given for at least four different purposes: to motivate students, to inform parents of their children's progress, to provide a record for employers and college admission officials, and to assess the performance of teachers and schools. This last purpose, which is needed for assessment and accountability of a public enterprise, doesn't need particularly sophisticated tests, since it is interested only in averages and standard deviations. Most of the problems associated with assessing individual performance vanish when class scores are averaged. On both the live test and the computer test, the average score of students who had had hands-on experience with batteries and bulbs was substantially higher than the average score of students who had had only textbook instruction, and both groups of students had about the same average on conventional achievement tests. These averages support the value of hands-on instruction over pure textbook learning, at least for batteries and bulbs, but no single test score characterizes an individual student's level of mastery with any great reliability.

Averaging is the most basic statistical technique, and the most important. It's taught in some seventh-grade classes now, and I was pleased recently to see students in a Cambridge middle school calculating the average, and even the range, of the data they had taken for their science fair projects. A SEED activity that quickly demonstrates the power of averaging has students, working in groups, measure the circumference and diameter of round objects of different sizes and write them on the blackboard. A wide range of values are displayed, but when each circumference is divided by its diameter, a number between 2.9 and 3.3 results. Averaging all these ratios magically cancels most of the individual differences that are due to random measuring errors, and a value close to pi ($\pi = 3.14\ldots$) emerges (Cromer and Zahopoulos, 1993).

Randomness is always a source of variation in any experiment, and contributes, in varying degrees, to the variations observed in test scores. The laws of probability allow one to predict how random variations are distributed. For example, imagine a ten-question true-false examination in which no student knew any of the answers. Each student chooses "true" or "false" randomly, and so, on the average expects to get half the answers "correct." Statistically, however, if 100 students take the test this way, only twenty-four will score exactly five, while thirty-eight will score above five and thirty-eight below. The five "brightest" students will get eights and and nines, while an equal number of "dummies" get ones and twos. Obviously, individual scores are meaningless in this case, and the class average of five indicates that the class was guessing.

The distribution of the scores on this test is shown as black bars in Fig. 9-1. This is technically called a "binomial distribution" and is the mother of all distributions. It's the actual count of the possible outcomes of a random process. For example, if a coin is tossed twice, the possible outcomes are heads-heads, heads-tails, tails-heads, and tails-tails, so 25 percent of the time one will get heads-heads. Such counting is necessary whenever the number of combinations is small.[5] When the number of outcomes is large, say the scores of 10,000 students randomly answering 100 true-false questions, the counting of all possibilities is neither convenient or necessary. The bar graph is approximated by a line graph, which passes through the top center of each bar. For purely random events, this line graph has a particular form called a "normal" distribution.[6] How this is related to the so-called bell curve will be discussed shortly. For now just bear in mind that binomial and normal distributions are theoretical predictions derived from the basic laws of probability. They apply only to random events.

As another example, suppose that on a different ten-item true-false test every student knows the correct answer to three questions and guesses on the remaining seven. Then the laws of probability can be used to predict the binomial distribution shown as white bars in Fig. 9-1. This is quite similar in gen-

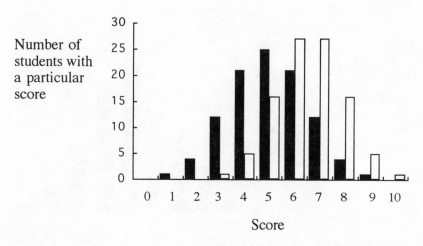

Figure 9-1 Distribution of scores on a ten-question true-false test. Black bars show the distribution of a class of 100 students, all of whom randomly guessed on all ten questions. White bars show the distribution of a class of 100 students, all of whom knew the correct answers to three questions and randomly guessed on seven questions.

eral appearance to the black-bar distribution except that the peak is between six and seven. The class average now is 6.5,[7] which indicates that some students knew some things. But if the tester thinks she has a "normally" distributed class, she may think some students know more than others, when, in fact, all the students knew just three answers. How can she know for sure?

Unfortunately, the only way to answer the question is to give the same poor kids another test on the same subject. If Joey got a nine on the first test and a five on the second, while Jane got six and nine, the tester should be a little suspicious. To systematically compare all her students' scores on both tests, she can calculate the correlation coefficient r between the two tests. This statistical test will yield zero in this case, meaning that all the variation in the scores is from random guessing.

Finally, imagine that half the class knows nothing and guesses on all ten questions, and half the class knows the correct answers to three questions and guesses on the other seven. The scores in this case will have the distribution shown in Fig. 9-2, which is the average of the black- and white-bar distributions in Fig. 9-1. Although it looks pretty much like the others, there is a real difference in performance buried here. To find it, the kids must be tested again, and the correlation coefficient between the tests calculated. If the same kids who knew nothing on the previous test know nothing on the second test, and if the kids who knew three answers on the previous test know three answers on this one, then $r = 0.26$. This indicates that some of the variation in the scores is not random: some students did know more than others. In this way, correlation studies can validate a test, that is, they can prove that a test measures real differences within a group, even though the test can't distinguish between smart high scorers who knew three answers and dumb high-scorers who were very lucky.

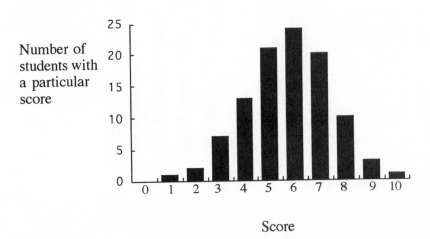

Figure 9-2 Distribution of scores on a ten-question true-false test on which fifty students randomly guessed on all ten questions and fifty students knew the correct answers to three questions and randomly guessed on seven questions.

Intelligence Tests

Nothing is more evident than that some people are smarter than others, and nothing is more vigorously denied. Differences in achievement are attributed to character, persistence, and socioeconomic status, but seldom to intelligence (Goleman, 1995). The Soviet Union banned intelligence testing of children on egalitarian principles, and the United States forbids the intelligence testing of prospective employees for fear of discrimination (Bowen, 1981; Herrnstein and Murray, 1994). Until I read *The Bell Curve* (Herrnstein and Murray, 1994), I thought, like everyone else, that intelligence testing was a long-discredited pseudoscience. After all, hadn't Stephen Gould told me so (Gould, 1981)?

The problem of measuring intelligence is similar to the problem scientists once faced in measuring temperature: how do you measure something you haven't yet clearly defined? And how can you define a fundamental concept in science, except in terms of still more fundamental concepts. The answer is that fundamental concepts, such as length, time, mass, and temperature, are defined by the procedures, or operations, used to measure them. Length is what is measured with a measuring rod and temperature is what is measured with a thermometer. Nothing more can be said about these concepts, and nothing more needs to be said. Words like "intelligence" and "temperature" have a number of everyday meanings, only a few of which are close to their scientific meanings. So there is often confusion between the technical and everyday meanings of scientific words, a price we pay for not using Latin.

Galileo built some of the first thermometers, and they were used in medical diagnosis long before the modern theories of heat and temperature were developed. This process of developing a fundamental concept from its method of measurement is an important part of the philosophy of science (Carnap, 1966). And because it's very concrete, it can be introduced in middle-school physical science (Cromer and Zahopoulos, 1993). Operationally, then, temperature is simply what we measure with a properly calibrated thermometer. But through a lengthy logico-empirical argument, this definition is ultimately related to the average energy of the molecules in a gas. Thus temperature, which is defined somewhat conventionally at first, has been found to be as basic a physical quantity as length, mass, and time.

The process of defining something we can call "intelligence" also involves inventing a procedure for measuring this something. As with temperature, the invention tries to capture some of the everyday meaning of intelligence, but in time the measuring procedure itself—the tests—define the technical meaning of intelligence. From this point on, the issue isn't whether the test measures *intelligence* in some absolute Platonic meaning of the word, but whether it reliably measures anything that can reasonably be called "intelligence." To avoid confusion, it's best to drop the word "intelligence" altogether, and just say we are trying to measure a cognitive attribute—call it *x,* or *g,* or IQ—that correlates with other quantifiable human behaviors. We really don't know what we are doing until we do it.

The statistical method of correlating scores on different tests to infer a cognitive difference in groups was developed by Charles Spearman at the beginning of the twentieth century. Objective, or multiple-choice tests were used, which, because of guessing, inevitably introduced some random shuffling of the scores. However, the magnitude of the random variation is calculable from the laws of probability, and is much smaller on a five-choice multiple-choice test than it is on a two-choices true-false test. All other methods of assessing human performance require some form of human judgment, which introduce noncalculable errors.

The scores on most IQ tests are designed to fit the normal distribution. This process is poorly understood. It's often thought that intelligence has been found empirically to have a normal distribution, and the book *The Bell Curve* does little to dispel this notion. Many variables, such as the term scores of students in my Physics 1 class—the sum of each student's quiz grades, mid-term grade, and final examination—have a "bell-shaped" distribution that is peaked near the average and that falls off rapidly on either side. A measure of the width of the distribution, called the "standard deviation," is calculated by subtracting each individual score from the average score, squaring these differences, summing the squares, dividing the sum by the number of scores in the distribution, and taking the square root of the quotient. For a normal distribution, 68 percent of the measurements are within one standard deviation of the average, 95.4 percent are within two standard deviations, 99.74 percent are within three standard deviations, and so on. For term scores in my Physics 1 class, 72 percent of the scores were within one standard deviation, and 100 percent were within two standard deviations. The distribution is bell-shaped, but not normal.

To achieve a more normal-like distribution, intelligence testers like to write tests in which there is no upper limit on the scores. This simply involves asking many more questions than anyone can answer. To get a broad distribution, these questions must be of varying difficulty. By definition, an easy question is one that almost everyone answers correctly. A difficult question is one that is answered by few people and these must be mostly high scorers. Test designers study this sort of thing, eliminating idiosyncratic questions and questions that have a regional or cultural bias. Over time, a mix of questions is found that yields a more or less normal distribution for two standard deviations around the peak.

The choice of a normal distribution is so ingrained that most social scientists have forgotten that it is so by design, not by nature, just as it is by design that there are 100 Celsius degrees between the melting temperature of ice and the boiling temperature of water. There are very good reasons for choosing ice, water, and 100, having to do with the availability of ice and water and the relation of 100 to our decimal number system. But as far as the concept of temperature is concerned, one could define the Cromer temperature scale in which there are 47.6 Cromer degrees between the melting and boiling temperatures of sulfur. It's a nasty choice to be sure, but not an illogical one.

Similarly, there are very good reasons for choosing to design tests to fit a normal distribution, the principal one being that, because of its use in other branches of science, tables of the distribution are readily available (Burt, 1963). These tables show that for a normal distribution—and only for a normal distribution—exactly 2.27 percent of the test takers will have a score that is more than two standard deviations above the average score. The test makers then design their tests so the scores, in fact, approximately distribute this way. If on one test they get too many high scorers, they make the next test a bit harder; if they get too many low scorers, they add a few more easy questions. In order to screen for very high scorers, the test has more questions than anyone can answer. By design, no one can get a perfect score.[8]

Test results are usually on a scale in which 100 is the average and fifteen is the standard deviation. A score of 130 is two standard deviations above the normal, which means that 2.3 percent of the test takers had scores above 130. A score of 145 is three standard deviations above the average, and 0.13 percent of a normal distribution is above three standard deviations. On a test given to a million people, approximately 1,300 should score above 145. However, since the fit of the scores to a normal distribution is always approximate, these percentages are also approximate. There are probably substantially more very high IQ individuals than is predicted by the normal distribution (Burt, 1963).

Having gone to all this trouble to design an instrument, one next must test its reliability. Does the thermometer always read the same under similar circumstances? Do two thermometers built to the same specification read the same temperature when placed together in a liquid? If not, the instrument isn't measuring anything. If so, the next question will be whether it measures anything interesting.

Over the years, a number of intelligence tests have been developed: Stanford-Binet, Wechsler, Kuhlman-Anderson, Lorge-Thorndike, Otis-Lennon, and so on. Correlations between these tests is typically 0.7 to 0.8 (Herrnstein and Murray, 1994). These are remarkably high, when we consider the inevitable random errors due to guessing on a multiple-choice exam. (The California sealed-boxes tests had a correlation of 0.9 for similarly trained students taking two tests designed by the same testers to be as identical as possible.) It's the high correlations that give the first indication that the tests are all measuring some common cognitive property.

The second indication of this property is the success researchers have in correlating test scores with other measures of human behavior: grades, income, marriage, and so on. In these cases, there is a positive correlation between scores and positive behavior. Herrnstein and Murray studied the negative correlations between scores and negative behavior. For example, they found that the probability of dropping out of school is much greater for low scorers than for high scorers (Chapter 4). This may not be too surprising; the tests could just be measuring a trait needed for schooling, such as reading comprehension. But they also found correlations, in an all white population, between low IQ and incarceration, unemployment, and out-of-wedlock babies.

The third indication that these tests are measuring something real comes from the correlations that have been found between the test scores of related individuals. The correlation between the test scores of identical twins, separated at birth, is a whopping 0.8, between siblings and between parent and child its 0.5, and between first cousins and between grandparent and grandchild its 0.25 (Dobzhansky, 1973; Jensen, 1969). Almost all these genetic studies have been done on whites, so it isn't known whether the same pattern of inheritance applies to other races. Nevertheless, the inheritability of IQ scores among one large racial group strongly supports the contention that the scores are measuring a substantial characteristic, since any inheritable trait has an underlying chemical basis.

Thus the evidence is quite strong that certain tests do measure an important attribute of human cognition. Because of the correlation of the scores on these tests with intellectual achievement, the attribute may properly be called "intelligence," remembering that in its technical context, intelligence means the attribute measured by certain tests, nothing more and nothing less. However, because *intelligence* is such an emotionally charged word right now, objectivity is better served by referring to the measure—the test scores—and not bothering to give a name to the attribute they measure.

Criticism of Intelligence

The Bell Curve stirred up a hornet's nest of anger and criticism (Jacoby and Glauberman, 1995), perhaps none fiercer than that of Stephen Gould (1994), a long-time foe of all attempts to quantify or reify intelligence. He is also one of my favorite authors, and like millions of his fans, I have deep respect for his judgment and wisdom. His revolutionary theory of punctuated equilibrium in evolution has received strong empirical support from a detailed study of bryozoan fossils (Kerr, 1995; Jackson and Cheetham, 1994). Nevertheless, on the issues raised by *The Bell Curve,* I find that his passion has caused him to blow out of all proportion what are, in fact, vanishingly small differences of opinion.

In a *New Yorker* article attacking the book, Gould openly admits his anger and indulges himself with an effluence of heated words and phrases: "anachronistic," "unprecedented ungenerosity," "pervasive disingenuousness," "almost willfully hiding," "conservative ideologues who rail," "grotesquely inadequate." It's painful to read Gould in such a mood. He flames the authors in one sentence, then parenthetically agrees with them in the next, and discounts his agreement in the third: "Virtually all the analysis rests on a single technique applied to a single set of data—probably done in one computer run. (I do agree that the authors have used the most appropriate technique and the best source of information. Still . . .)."

Gould is most angry over the book's failure to justify that what it calls intelligence is "a real property in the head." This is similar in some respects to Mach's objections, on philosophical grounds, to the introduction of invisible atoms into science (Chapter 2). In Mach's time, the evidence for atoms was circumstantial, just as is the evidence for intelligence today. The justification that IQ scores are a measure of something we can reasonably call "intelligence" comes from its genetic and behavioral correlations—it isn't anything we can see or touch. On the other hand, intelligence is a far more intuitive concept than is the concept of atoms. Indeed, it's the denial of intelligence as a real human attribute that is counterintuitive.

Gould has a more serious technical criticism of the book, which concerns the relatively low values of the correlational coefficients r found between IQ scores and negative social behaviors. Gould writes that "Herrnstein and Murray actually admit as much in one critical passage, but then they hide the pattern." This seems a bit unfair, since Appendix IV of *The Bell Curve* has twenty-eight pages of tables containing the results of their statistical analyses, including the square of the correlation coefficient r^2 for each analysis. The square of the correlation coefficient is often used as a measure of how much of the variation in a sample is produced by the independent variable. However, the theoretical basis for this interpretation of r^2 applies only to situations in which a plot of the dependent variable against the independent variable is, more or less, a straight line. This wasn't the case for the income versus education data plotted in Chapter 4, and is even less so for the negative social behavior versus IQ data. For this reason, Herrnstein and Murray discount the significance of their low r^2 values, relegating them to the appendix.[9]

More telling is Gould's criticism of the characterization of intelligence by a single number. Surely it makes no more sense to add together scores from verbal and mathematical tests than it does to add together apple and pears. But, contrary to common opinion, you can add apples and pears: ten apples and seven pears is seventeen pieces of fruit. Furthermore, if it's given that a bowl can have no more than ten fruits of the same kind, then I know that a bowl with nineteen fruits has either ten apples and nine pears or nine apples and ten pears, whereas a bowl with only five fruits can have at most five apples or five pears. Bowls with a large number of fruits are very distinguishable from bowls with a small number of fruits: a bowl with five fruits has many fewer apples and many fewer pears than a bowl with nineteen fruits. Bowls with a moderate number of fruits can be very different from one another: One bowl could have nine apples and one pear whereas another could have one apple and nine pears.

People with average IQs have their strengths and weaknesses, whereas people with high IQ have mostly (cognitive) strengths and people with low IQs have mostly weaknesses. This is so uncontroversial that even Gould says the same thing:

> Of course, we cannot all be rocket scientists or brain surgeons, but those who can't might be rock musicians or professional athletes (and gain far more social prestige and salary thereby), while others indeed serve by standing and waiting. (1994)

I don't know how to interpret this except as an admission of the general proposition that human beings vary in intellectual capacity. Yet in the same paragraph he says we must fight the doctrine that there is "a single scale of general capacity with large numbers of custodial incompetents at the bottom."

The Bell Curve views with alarm the transformation of the United States into a custodial society. This is the conservative perspective. Gould views with alarm recognizing that there are people of low intelligence who need some custodial care. This is the liberal perspective. In fact, the United States and all other developed countries are custodial states. This is the realist perspective. It's not just the five million people in prison or on parole and probation, nor the millions in public housing and mental institutions. It's also the fifty million young people who are forced by the state to creep snailward to school each day, and those adults in the assisted-living and rehabilitation facilities that I pass each day on my five-minute walk from my home to my bagel shop.

First there is the public school building that has been converted into housing for the formerly homeless. This is custodial in that it has resident counselors who see that the residents take their medications and attend their AA meetings. This is real stuff. One of the residents used to sleep in the doorways near the bagel shop; now he's a customer. Next to the school is the Boston Center for Independent Living, a custodial residence for quadraplegics. These brave souls careen down the street in their breath-control, battery-operated wheel chairs, as heart-stopping a sight as any I've seen. Then, as I reach the corner of Centre Street, I pass Crossroads, a day facility for the mentally ill. Across the street, the bagel shop caters to us all.[10]

Conservatives don't want to know about these people and liberals don't want to believe that there are people who are truly different from them.

Politics of Special Education

Of course, the main criticism of the *The Bell Curve* isn't about the custodial state or r^2 values, but is about race. Because many studies have found that African-American score, on the average, about fifteen points lower on intelligence tests (Shurey, 1966), any talk about intelligence is felt to be both insulting and injurious to them. Many educators, both black and white, denounce grouping students by ability as a form of *de facto* segregation. In a 1992 law suit, the NAACP called ability grouping "one of the most divisive and damaging school practices in existence" (Caldwell, 1992). This antigrouping position is threatened by the acceptance of intelligence as an important cognitive parameter because such acceptance gives theoretical support to the position that students should be separated by ability.

The antigrouping position is ideological, rather than political, because it doesn't obviously benefit its own constituency. Just as strong arguments—often by the same people—are made for maintaining voluntarily segregated dormitories on "integrated" university campuses and for not closing some historically black colleges to promote integration. Ideology is seldom consistent enough to be the basis of rational policy. In this case, it's too busy fighting Jim Crow and Adolf Hitler to notice that other interest groups have successfully lobbied to get extra funding and classrooms for their children.

The special-education movement is political, not ideological, since it's about taking money away from one group of children and giving it to another. Public schools in the United States spend 50 percent more on a child declared "learning-disabled" than on a child who isn't so labeled. The consequence, not surprising, is an epidemic of learning-disabled children, jumping from 783,000 in 1976 to 2.3 million only seventeen years later (Roush, 1995). New York city alone spends $200 to $400 million a year just testing students for learning disabilities, a totally unvalidated process that is performed by an outside army of 2,800 psychologists and educational evaluators who receive over $2,000 for each evaluation—all under the watchful eye of 2,000 central and district special-education administrators (Dillon, 1994).

This all started when parents' groups succeeded in pressuring Congress into legally recognizing difficulty in learning as a disability under the Individuals with Disabilities Education Act, thus entitling slow learners to the special services warranted for children with speech impairment, mental retardation, and physical handicaps. But learning disability is a vague concept, to say the least, and parents who understood this fought to get the disability tag put on their children because of the special attention to which it entitled them. As one upper-middle-class mother said about her college-bound children: "If it hadn't been for special education, they would have been just average."

We thus have the ironic situation that one group of parents—largely white and middle class—doesn't object to their children being labeled "slow" if that gets them the extra attention they need, while another group—largely black and poor—actively fights labels that they feel will stigmatize their children. Both groups claim the purest motives, although one group has its hands in the other group's pockets. The percentage of learning-disabled students in a state largely depends on the resources available to meet their needs. It varies from under 3 percent in Georgia to over 9 percent in Massachusetts (Roush, 1995).

The current definition of a learning disability is a reading level below that expected on the basis of tested or inferred IQ and which can't be accounted for by other physical or emotional problems. It is a bit of a catch-all, which has prompted calls for more research. In response, the National Institute of Child Health and Human Development has spent a miniscule three million dollars a year on the problem, which is one percent of what New York City spends on diagnostic testing. Nevertheless, some intriguing results have been reported, among them evidence that, while interpreting letters, there is less activity in the left thalamus of a poor reader than in that of a good one. This by itself doesn't prove very much, since it isn't clear whether the thalamic activity is low because the poor reader isn't reading or the poor reader isn't reading because the thalamic activity is low. Other studies have shown that reading disabilities run in families, with a heritability of between 50 to 70 percent, and Cardon et al. have traced the trait to a specific region on chromosome 6 (1994).

This last result, if it holds up, is extremely important, since it maps a complex behavioral trait to a specific site in the genome. It supports the contention that learning disability—a contributer to low test scores—is a real biological trait. It's thus conceivable that there could someday be a genetic test for dyslexia, a thought that not everyone in the learning-disability community is comfortable with. Psychologists could no longer demand $2,000 for ambiguous tests, and some children now classified as dyslexic would be found not to be, and vice versa. Furthermore, this same scientific research is showing that 20 percent of the population may have the trait, which would make it not so special after all. The more widespread the trait is, the smaller the justification is for spending more on the education of children with the trait than on the education of children without it.

Politics of Intelligence

It's useful to imagine what the arguments would be if intelligence was treated politically, rather than ideologically. In this case, the focus would be on the special needs of the population with low intelligence, since the 10 percent of the population with IQ scores below eighty-one have problems that are substantially different from the rest of the population. All the social ills—truancy, poverty, crime, violence, alcoholism, drug abuse—increase rapidly as IQ falls below eighty. As one inmate in a Texas prison drug rehabilitation program put it : "It ain't that we're doomed. . . . But we have the real potential to mess up. One thing leads back to another" (Verhovek, 1995). In simpler times, such people mostly worked as farm laborers, living as poorly as everyone else. They had fewer opportunities to "mess up," and little was expected of them, except to live and die as they were told.

To avoid having to repeatedly use such highly charged terms as "intelligence" and "IQ," I will mostly speak of level I, II, and III people. Level I people are those who, for whatever reasons, have excessive difficulty in school and in life in general. This groups includes not only low-IQ people, but the lazy and alienated as well. By definition, most of these people are poor, but not all poor people are in level I. Level II people are mostly in the broad middle-range of IQ, but in level II there are some hard-working people with low IQ and some lazy people with high IQ. Level III people—the gifted and talented— have both high IQ and ambition. A politics of intelligence is one that recognizes the different needs of these populations.

Modern urban society expects everyone to live autonomous lives and to be skilled in a number of complex tasks, such as finding employment, managing time, controlling money, paying bills, shopping for price, regulating sexual activity, avoiding addiction, and so on. Yet a recent study conducted for the U.S. Department of Education by the Educational Testing Service found that a fifth of the adult population in the United States couldn't total a bill, determine a price difference, or fill out a form. Two-fifths of the population couldn't answer questions about a newspaper story they were given to read or write a paragraph summarizing information on a chart. Only a quarter of the population were considered highly literate (Celis, 1993).

The poor performance of 40 percent of the population is probably a reflection of how badly our schools serve level I and II children. From my experience with college students, I have concluded that even many level II and III students fail to make the transition from concrete to formal thinking because their teachers never made the transition (Cromer, 1993). From more recent experience with prison inmates, I am finding that level I students have a more literal understanding of words than do average students. They can properly connect a simple electrical circuit, but can't answer questions about the path of the electrons. The problem isn't with the word "electron," which they accept as something moving in the wire, but with the word "path." They seem unable to generalize the word "path" from its everyday meaning as a walkway to its scientific meaning as a continuous route from here to there. It's certainly clear that these men have educational needs that can't be met in a classroom filled with average and above average students.

Thus a politics of intelligence would insist that level I students have instruction appropriate to their needs. They must master preschool skills that involve counting objects, recognizing shapes, and manipulating small objects before they go on to beginning reading and arithmetic, even if this means delaying their reading for a year or so. At each stage, there must be enough repetition so that a firm basis is made for the next step. Reading should be related to manipulable objects, so that the meanings of the words will have a concrete reference. Special attention should be paid to simple words, like the conjunctions "and," "but," and "or," which may be used inconsistently by level I students.

Students would be ability-grouped on the basis of their attainment in preschool and kindergarten, not on the basis of an IQ test. There must be regular review of students' progress, and students transferred between groups as their progress dictates. The concept of intelligence plays only a theoretical role in all of this. It says that there are students with a need for slower classes, but it doesn't say who they are. They are identified, after the fact, as those who benefited from being in such classes.

A politics of intelligence would insist on policies that encourage level III families to stay in urban areas. This means that urban schools must meet the special needs of high, as well as low, IQ students. Cries of "elitism" must not prevent schools from offering challenging programs for the brightest students, because such programs help to control the draining of brains from both white and black communities. The venerable Boston Latin High School is a case in point, still requiring Latin after all these years. Far from dividing the population, private-school students compete for a place in this elite urban public school.

A politics of intelligence would insist that level I people get the help they need to cope with the complexities of modern life. Surprisingly, most can work; a full 78 percent of white men in the lowest 5 percent of intelligence (IQs below 75) were employed in 1989 (Herrnstein and Murray, 1994). This is a tribute to the low-wage policy of the United States, which greatly increases the employment opportunities for low-skill workers. Europe, with a high-wage policy, suffers twice as much unemployment as the United States. Thus, raising the minimum wage in the United States would probably hurt the level I population the most, just as conservatives claim. But, by the same token, it is disingenuous to expect this population to function with pure Calvinist probity without considerable social support. They need simple nonprofit financial and personal services, from check cashing to advice on budgeting. It's one thing to pay them low wages, it's another to charge them for cashing their checks.

Most important, a politics of intelligence would insist that level I people get the vital information they need in a form they can understand. This is a difficult challenge, because the producers of the information don't understand how these consumers think. But judging from the large amount of miscommunication that goes on among high-IQ types, probably none of the messages about nutrition, AIDs, drugs, pregnancy prevention, and so on reach people with the lowest IQs. To be effective, they probably should be overly simplistic. Leave mixed messages like "Abstinence is best, but if you choose to have sex, always use a condom" for the Boston Latin crowd.

A politics of intelligence would insist that the cognitive differences among human beings be recognized, and that people be treated appropriately. This necessarily means ability grouping in school. Such grouping cuts across class and racial lines, though not proportionally. Given the current distribution of IQ in the white and black populations, there will be a higher percentage of black students in level I classes than there are in level III classes, but all classes will be integrated to some degree. No rational definition of integration requires that each classroom have some predetermined proportion of races in it, because there is no rational way to decide such a proportion. Should it be that of the students in each school, in each district's public schools, in each district as a whole, or in an entire region? One of the purposes of ability grouping is to redress the racial imbalance of urban public education that heterogeneous grouping has produced.

It doesn't take an exceptionally high IQ to recognize that heterogeneous grouping with meaningful standards is impossible, because it would fail too many students. But an enormous amount of creative energy is wasted in denying this impossibility. Contradictions pile upon contradictions. *National Science Education Standards* calls for all students reaching a defined level of understanding and ability, while admitting students will achieve these outcomes at different rates (National Research Council, 1996). Since it will take some students longer than others to achieve the "standards," I fail to understand how they can all be in the same single-age class.

> Educationalists have always shown to an extreme degree the besetting sin of many people who wish to change the world; the refusal to acknowledge the diversity of human nature. . . . The notion that there is one best method of teaching, one single method which is applicable to all children regardless of personality, ability or circumstances, dies hard; recognition that the very notion is an insult to human individuality seems a necessary preliminary to any improvement in our system of teaching. I can see nothing wrong in adapting teaching methods to human differences. (Eysenck, 1971)

The politics of intelligence doesn't try to eliminate human differences by denial, but insists that these differences be considered in making public policy. Herrnstein and Murray fear that any advances in education and learning technology will disproportionately benefit the more intelligent, who can make better use of them. This might be true if students were currently achieving at anywhere near their potential, but they aren't. This is especially true of level I students, who, because they are more dependent on the skills of their teachers than are level III students, are disproportionately injured by poor teaching. Unfortunately, politics and bureaucracy have broken the morale of many urban teachers, making improvements there maddeningly difficult to effect. In Boston, precious professional development time has been wasted on senseless and impractical workshops, such as how to produce systemic change, how to teach science by making ice cream, and, worst of all, how to understand *National Science Education Standards.*

My own work has been directed toward workshops that develop teachers' understanding of basic physical concepts through hands-on activities they can use with their students. The teachers have been enthusiastic about this approach and, since 1995, our workshops have been integrated into state and urban professional development programs. I've also developed and taught a hands-on science program in Boston's most costly school, the Suffolk County House of Corrections. Prisons may be the perfect place to study ways to best teach level I students, because the tight security and rigid discipline eliminates most of the students' disruptive behavior.

Progressives, always ready to blame people's problems on their environment, have failed to recognize that level I students may need a different learning environment than other students need. Conservatives, always ready to put a Calvinist spin on everything, are eager to blame the problems of the poor on their immoral behavior. Perhaps a more balanced approach can be achieved by recognizing the role that human differences—particularly differences in intelligence—play in people's lives and adopting policies that address these differences in a constructive way.[11]

Optimism begins by recognizing the enormous inefficiency of the U.S. educational system. Indeed, the word "system" is an unfortunate euphemism that disguises the disarray and incoherence of instruction from middle school through college. By comparison to education in other Western countries, education in America is simply shameful. Thus, with apparently no where to go but up, there is hope that improvements in education can have major benefits for students of all intelligence levels.

How have we come to such a state? Why is U.S. education so different from that of all other countries? And most important, what can be done about it? Is there a magic bullet, an off-the-shelf, zero-cost solution to all our problems? The analysis in the next chapter suggests that there is, and that it comes, surprisingly, by thinking about the needs of level I students.

10

Making Connections

The United States considers itself to be a new country, especially in comparison to the ancient civilizations of Greece and Italy. Yet in their present constitutional incarnations, most of the nations of Europe, including Greece, are younger than the United States. Greece lost its independence to Rome in 146 B.C. and regained it from the Turks only in 1832. The Greek constitution was written in 1844, fifty-five years after the U.S. constitution.

It wasn't as clear in the eighteenth century as it was in the nineteenth century that education is just as fundamental a function of national government as is the military and the currency. As a consequence, the U.S. Constitution didn't give Congress any authority over education. Americans still view education as primarily a local responsibility and no politician or governmental official ever argues otherwise. There are 16,000 independent school districts in the United States, only one of which—New York City—is the size of the Greek school system. The average U.S. school district has only 3,000 students in it, so in spite of the size and wealth of the United States, few districts have the resources to develop coherent and comprehensive curricula. Without such curricula, individual teachers are left to their own devices. Even in the same building, different sixth-grade classes will be studying different topics which neither follow upon what was learned in fifth grade nor prepare for seventh grade. This haphazard nonsystem produces critical gaps in student preparation which are most damaging to the least able.

In other countries, education is controlled at the national level by the publishing of the textbooks. There is one book for each course at each grade level. Even in Greece, with a per capita income less than one-third that of the United States, every student receives a new book in each subject every year. These books are softcovered, printed in color on noncoated paper, and contain just the material to be covered during the year, no more and no less. By seventh grade, all courses are taught by teachers with the equivalent of a master's degree in the subject. Currently, all Greek seventh graders take the following fifteen courses:

Modern Greek
Mathematics (algebra and geometry)
Modern Greek literature
Computer programming
Ancient Greek
Biology (botany)
Human anatomy
Ancient Greek literature (in modern Greek: Homer's *Odyssey,* Herodotus'
  Histories, and Xenophon's *March of the Ten Thousand*)
Religion
English
Art
History (ancient civilizations to the conquest of Greece in 146 B.C.)
Music
Geography
Gymnastics

The depth and scope of the Greek seventh-grade curriculum is daunting by U.S. standards, and in eighth grade, physics and chemistry are added to it. Serious students spend many hours a day on their homework and high anxiety is considered normal. Greek teachers traditionally know little, and care less, about fostering their students' self esteem. Most students leave after nine years of school, but with a far stronger education than American students have after twelve.

All Greek students, public and private, use the same books in the same subjects, although private-school students may study German or French as well. With fewer pages, a softcover, and uncoated paper, the books are much lighter than comparable American textbooks, which is necessary because the Greek student carries a dozen of them back and forth to school. The contrast couldn't be starker with the United States, where it is common, even in wealthy school districts, for there to be a shortage of books for in-class use, let alone for students to take home with them. Indeed, students' having textbooks in the United States is becoming so rare that it makes the newspaper. An article in the *Providence Journal Bulletin* described how the students in one English class became more attentive and improved their work after they got their textbooks in February (Frank, 1995).

In Greece, the students receive all their textbooks at the beginning of the school year so that they know, their parents know, and their younger siblings know exactly what is studied in seventh grade. Since the students own their books, even the poorest family accumulates books as their children progress in school. The whole society is conscious of what is taught at each grade level, which creates a uniformly high level of expectation. In the United States, teachers, untrained in subject matter, have almost complete freedom to devise their own curricula. A Montana school teacher received the 1995 Presidential Award for Excellence in Science and Mathematics for devoting her entire science curriculum to the study of spiders (Potera, 1995). This degree of teacher autonomy is the *reductio ad absurdum* of local control, which is too weak to develop and enforce a coherent uniform program of study.

The current drive for state and national standards acknowledges the need for some uniformity of curriculum, but the documents that have been produced to date are too diffuse and ambiguous to be of much help to individual districts in the difficult task of writing curricula. Topics are listed without any suggestion as to the order, or even the year, in which they are to be taught. For example, *Massachusetts Science and Technology Curriculum Frameworks* says that by the end of eighth grade, students must be able to "Compare and measure different materials in terms of their characteristic properties such as density, texture, color" (Massachusetts, 1995). Density is, in fact, the only characteristic property that can be measured in middle-school science. Listing it along with texture and color, which aren't easily quantified, is misleading at best.[1] Density is the ratio of two different measurable quantities (mass and volume), so prior knowledge of the measurements of these quantities is needed. Furthermore, ratio is a sophisticated concept that must be developed through a sequence of simpler activities, such as measuring the ratio of the circumference to the diameter of a circle. This ratio, pi (π), is important because it's independent of the size of the circle, just as the density of a piece of clay is independent of the mass of the clay.

Density should be a key topic in any middle-school science curriculum, but its proper treatment requires the development, over many months, of many important antecedent topics. Ideally, students should do experiments to measure the densities of solids, liquids, and gases and, subsequently, density should be incorporated into their study of pressure (see Table 1-1, Chapter 1). The Massachusetts Department of Education, in spite of the millions of federal dollars it received for science reform, made no attempt, even as an academic exercise, to show where density, or any other topics in its own framework, fits into a reasonable middle-school curriculum. Individual school districts, with their limited resources, are expected to do what in other countries is done by national governments.

But don't we need some sort of uniformity and coherence to our educational programs? Isn't that what I'm arguing for? Yes, it is. But we already have a national standard. Actually, it's an international standard, since it's recognized throughout Canada as well as the United States. It has been in existence for over fifty years, and it's consciously ignored by all public school officials. Therefore, my first off-the-shelf, zero-cost suggestion is that state and national organizations stop duplicating each others' efforts writing new standards and adopt the standard we already have: the tests of General Educational Development (GED tests).

GED Tests

The GED tests are nationally administered examinations for adults who have not graduated high school. Passage of these examinations is recognized by all states and provinces as the equivalent of a high school degree and provides qualification for admission to college and to thousands of postsecondary certification programs. Almost all avenues of advancement, except perhaps boxing, are closed to those without a high school diploma or a GED certificate.

The GED originated during World War II, when most of the draftees hadn't graduated high school. Since 1971, over six million adults have received a GED certificate through the GED test program, now administered by the American Council on Education at 2,900 testing centers throughout the United States and Canada. Currently, about 500,000 adults receive GED certificates each year, which is about 20 percent of the number of high school diplomas granted (Barasch et al., 1993).

My second off-the-shelf, zero-cost recommendation is that the GED tests be given to all ninth graders. In many states, these examinations would replace some other statewide test at this level, and so wouldn't be an extra expense. Currently, many states are spending, or planning to spend, millions of dollars to develop their own tests, instead of utilizing the existing national standard.

Current policy restricts the GED examination to adults, in order that it not compete directly with the K–12 school system. This isn't in the public interest. If a GED certificate is needed to go to barbering school, it certainly should be required to continue beyond ninth grade. Students who pass the GED tests in ninth grade should be given their GED certificates in a grand graduation ceremony. Only they, and perhaps students who passed four of the five GED tests, should be permitted to continue their education in high school, college, or professional training programs. All others should stay in eighth and ninth grades, as appropriate, and take the examinations again the following year. Even the students who missed only one examination (which will almost always be the Mathematics test), must continue to study for this examination even as they take more advanced subjects in other fields.

In short, the GED tests should become the universal minimal standard for all students. There are two possible objections to this. The first is that the standard is too low. Isn't the GED some consolation prize for dummies? No. The newest version of the GED tests, introduced in 1988, consists of five examinations—Writing Skills, Social Studies, Science, Interpreting Literature and the Arts, and Mathematics—that contain a total of 286 multiple-choice questions and one essay and are to be completed in seven hours and thirty-five minutes. The emphasis is on interpretive and evaluative skills, and not specific facts. Most questions relate to a diagram, drawing, or reading. For example, after reading a paragraph about a study done after the accident at the Three Mile Island nuclear generating plant, the examinee is asked to infer what had happened at the plant. From the study of adult literacy reported in Chapter 9, we know that 40 percent of the population can't answer questions of this type. And in college, students now are being taught how to use the conjunctions "and," "or," and "but" (Dembner, 1995). The GED mathematics test is even more challenging, involving interpretation of graphs, calculation of probabilities, fractions, percents, and simple algebra and geometry. Many college students, even those taking calculus, have no facility with this type of basic computational arithmetic.

This brings up the alternative objection, that the GED tests are too challenging to be the gateway to further education. After all, some students will fail
them. Don't we want everyone to succeed? On this point, even *National Science
Education Standards* and Goal 2000 people concede that although everyone is
to succeed, and there is to be no ability grouping, not everyone will succeed at
the same rate. Overlooking the slight inconsistencies in the logic of our educational leaders, they do seem to accept, in their indirect and obscure way, that
some students will be held back. Any standard that isn't totally meaningless
must have students who fail to meet it. The only valid question is whether the
standard realistically assesses the readiness of an individual to continue with his
or her education. That is, when we require a student to repeat a course of study
preparatory to the GED tests, are we absolutely certain that this is in the student's best interest?

I think that a strong case can be made that the GED tests assess, as well as
any finite set of tests can, the minimal skills that are necessary for almost all
education and training beyond ninth grade. The level is still light years below
that of a Greek seventh grader, but at least the GED tests put a floor under our
feet.

The standard for passing the GED tests varies slightly from state to state, but
nationwide 63 percent of those taking the tests in 1994 did receive their equivalency credentials. This is quite remarkable, considering that the population
taking the tests is composed largely of school dropouts, who, on the average,
have below average IQs (Chapter 4). It's estimated from tests conducted in
1987 and 1988 that only 70 percent of high school graduates could pass the
GED tests (Baldwin, 1995). Since the adults who take the GED tests have, on
the average, a tenth-grade education, it would seem that the last two years of
high school have done very little to bring the bottom third of high school students up to the minimum GED standard.[2]

Most students could be brought to GED standards by ninth or tenth grade
if schools focused their curricula on this task. Curriculum development is no
easy task, but here too there is a zero-cost, off-the-shelf solution.

Middle-School Curriculum

A common objection to developing a curriculum to meet a standard assessed
by a multiple-choice test is that the curriculum will "teach to the test," that is,
that it will focus on a narrow range of material. This objection has some validity when the test is a collection of specific facts, because then there may be a
temptation to teach facts rather than understanding. But when the test is on
general reading, writing, and computational skills, one can hardly object to a
curriculum designed to develop these skills. These skills aren't a mystery. They
can be broken down into their component parts, such as understanding the
meaning of words like "but," "however," "even," and "only," and understanding
the relation between fractions and decimals.

The off-the-shelf, zero-cost way to develop a sixth through ninth grade curriculum to meet the GED standards is to adopt the books currently being used for Adult Basic Education (ABE). ABE is a whole other educational world, a parallel universe. It generally isn't under the jurisdiction of the local school board, its teachers may not have degrees in education, and its books aren't written and published by the authors and publishers of standard school books. Subject more to market than bureaucratic forces, the ABE publishers have produced a collection of inexpensive books that focus on basic GED skills.

Seymour Rosen has written an excellent set of twelve inexpensive middle-school science books that, although not specially for the GED examinations, have much of the same focus on skill building. To appeal to students reading below their grade level, the books all have a 4–5 grade reading level and a 6-to-adult interest level. I became familiar with this series when I assisted in a low-level ABE class for prison inmates that used Rosen's *Electricity and Magnetism* (1994). Each chapter in a Rosen book contains only a page or two of text followed by four or five pages of simple questions. However, within this very simple format, it covers real science in remarkable detail. For example, *Electricity and Magnetism* goes into electrical transformers, discussing the primary and secondary coils and how a step-up transformer that increases voltage from 100 to 200 volts must have twice as many turns on the secondary as on the primary coil. This is real-life science which one just doesn't find in standard middle-school books.

My role in the ABE class was to assist the teacher in the class with demonstrations, experiments, and color commentary. I asked the class where transformers are found and passed around a transformer taken from a low-voltage AC/DC adaptor used to power small electronic appliances. The students demonstrated their understanding of the chapter on transformers by explaining the meaning of "PRI 120 V" and "SEC 12.6 V" that were printed on the transformer, by recognizing that this was a step-down transformer, and figuring out that the primary coil must have ten times as many turns as the secondary.

The point of this and other lessons wasn't so much to teach isolated facts about science, but to enhance basic reading and comprehension skills. There was a lot of emphasis on dichotomies, such as primary and secondary, step-up and step-down, and the relationships between them. The questions, much simpler than on the GED examination, repeatedly drilled away at these relationships: "In a step-up transformer the secondary coil has (more, fewer) turns than the primary coil."

This book is focused in purpose and execution, giving it the simplicity and utility of a paper clip. The readings discuss real-world science in simple, but nonpatronizing, language, while the questions develop understanding of the relevant relational terms—greater–lesser, more–less, thicker–thinner. The book provides a beautiful blend of language and cognition, especially when time is taken to do the experiments it describes.

This unpretentious single-author work, printed on uncoated paper with one accent color, is in stark contrast to the glitzy, hardcover, full-color works published for the school market. Developed by an army of writers, review boards, and field testers, they are so busy pandering to all the latest educational fads that they lose coherence and comprehensibility. Chapter 1 discussed one particularly sad example: *Middle School Science & Technology* (BSCS, 1994a, 1994b, 1994c). The three books in this series—*Investigating Patterns of Change, Investigating Diversity and Limits,* and *Investigating Systems and Change*—are each organized around a few unifying themes, such as change, diversity, and limits. Unfortunately, these themes aren't the way science organizes its understanding of the natural world. The disembodied concept of change as an organizing principle went out with Aristotle.

Almost none of the activities and investigations in these books use real-world methods to deal with real-world science. For example, one activity involves listening to the sound of popcorn popping. "Squat as low as you can beside your desk. . . . Indicate the amount of popcorn popping at one time by adjusting your height according to the sounds of the popping" (BSCS, 1994b). Then the students are instructed to make a graph—with their eyes closed. The point of this, believe it or not, is to demonstrate diversity. Each student, you see, will draw a very different graph. The students aren't asked to suggest ways to decrease the diversity. The concepts of objectivity, measurement, and repeatability aren't being suggested here, the task being hardly appropriate for this. Instead, the students are asked to describe the amount of diversity in their classmates' drawings as either a lot, some, or none. In terms of reading level and concept development, this "standard" seventh-grade book is far below that of Rosen's "remedial" works.

Things don't improve in eighth grade. By page fifteen of the eighth-grade book, the concepts of dynamic and stable equilibrium have been hopelessly mixed up. On page seventy-three, we learn that "Acids and bases also are types of chemicals. In concentrated form acids and bases are very harsh. . . . Strong acids have many hydrogen particles (H) present" (BSCS, 1994c). What is hydrogen? What is a hydrogen particle? This is their first and last mention in the book. The ancient Chinese and Greek understandings of elements are discussed 130 pages later, but poor Lavoisier and Mendeleev, and all of chemistry since 500 B.C., are nowhere to be found.

This series consciously avoids developing concepts in any orderly fashion, with the result that students are left in a muddle. They are told that an antibody is a protein, but not what a protein is. Terms like "atom" and "molecule" aren't used, whereas terms like "hydrogen sulfide" are. In spite of extensive field testing, the books are filled with factual and pedagogical errors. It's as though no one, including the distinguished people on the developer's advisory panel, ever read these books. As outlandish as this claim sounds, I know for a fact that one NSF-funded center that reviews new science curricular materials was going to put the BSCS series on its approved list without reading it. When I objected, the center's staff graciously reviewed my criticisms before giving the series its approval. It wasn't that they disagreed with my judgments, but they felt they couldn't reject new curricular material that was itself developed with National Science Foundation funding.

In the 1960s, the NSF sponsored a number of very successful curricula that were developed, for the most part, by university faculty. In its current reincarnation, the NSF's effort at curriculum development is being done by teachers and educators who have no expertise in the subjects they are writing about and who, for the most part, are wholly dependent on an NSF grant for their income. This lack of financial and intellectual independence has created a network of insecure educators who support one anothers' claims of authority in science education and who turn their backs on academic scientists. Constructivism, with its denigration of objective knowledge, has been the theoretical justification for this institutionalization of ignorance.

The failure of NSF-sponsored programs to develop focused and coherent educational materials results from the constructivist belief that education should start from the complex and work toward the simple. Thus, instead of systematically developing the concept of pressure, it's to be introduced when teaching about hurricanes or the gas law. But the idea that the explanation for things can be intuited from their surface appearances has defeated every empiricist since Aristotle. The principles of pressure aren't inferred from a study of hurricanes—hurricanes are understood from a study of the principles. These principles can be taught using real-life, hands-on experiments that are totally within a student's control. This approach, from the simple to the complex, may be tedious at times, but there is no other way to meaningfully build the connections between theory and observation. And it's these connections, after all, that are science.

There is little hope that middle-school curricula based on this approach will be developed nationally in the foreseeable future, so school districts interested in realistic educational reform must look elsewhere. I believe excellent middle-school programs, designed to prepare students to pass the GED tests by ninth grade, can be assembled from the existing stock of GED and specialty books. In general, these have a focus and coherence not found in mainstream textbooks.

Custom Publishing

Besides focus and coherence, school districts must have greater uniformity in their curricula. All students in each middle-school grade in all schools within a district should own and use the same set of basic books. This uniformity is essential for a number of reasons. First, it assures all parents that all children are working toward the same minimal goal of passing the GED tests by ninth grade. Second, the books are the means by which students achieve this goal. Ownership of their school books is especially important for disadvantaged students, whose homes may not otherwise contain many books. Third, with a common curricular core, teachers in a district can learn from one another ways to improve their teaching. Not many science teachers may think of bringing to class a transformer from a low-voltage AC/DC adaptor, but once one does, all other teachers using the same book may also want to bring one to class. Fourth, uniformity enables the schools to have curricula that progress from grade to grade without gaps or unnecessary repetitions. Fifth, by providing each student his or her own books, a school district creates a market large enough for custom publishing.

Custom publishing, common at the university level, is the professional printing of books for a specific course. Usually the material is written by the professor of the course, but this isn't always the case. Nor does the material have to be in camera-ready format. Some custom publishers will do all the word-processing work, including creating the line drawings and scanning photographs. In quantities of 500 copies or more, the cost for a two-hundred-page book is about ten dollars a copy. The publisher makes money by the annual reorders. This is the essential economy of custom publishing. By ordering ten-dollar-books yearly, instead of thirty-five-dollar-books every five years, schools can afford to give every student textbooks tailored to the district's curriculum.

In most cases, districts wouldn't write their own material, but would create their customized books from existing copyrighted material. For example, a district might create an eighth-grade science curriculum around material from several of Rosen's books. His publisher would then produce a customized book with just the materials that this one district selected. Another district might want a different selection. Digital technology makes it economical to do such cut-and-paste jobs for relatively small orders, especially if the orders are repeated annually. The author's royalty would, of course, be included in the price.

Custom publishing allows for annual revisions, so the district isn't locked into a fixed program. These revisions can be anything from the rewording of sentences to the substitution of whole topics. The original authors might be asked to add a specific subject. Public domain materials, such as NASA photographs, could be added when appropriate. Also, the authors of out-of-print books often give permission to reprint sections of their works for little or no payment. With modern digital technology, schools have almost unlimited possibilities for creating their own curricular materials with limited resources.

As a step in this direction, I have been working with the Fall River Public Schools on a custom book for a new sixth-grade science curriculum. The sixth-grade teachers and I meet once a month to discuss the contents of the curriculum and to conduct suggested activities and experiments. Much of the material came originally from the SEED program, but new ideas have been developed for investigations in biology and microscopy. Several principles are emerging from this work, which may be of interest to other curriculum developers:

1. The curriculum is concept driven. Concepts are developed in logical sequence, starting with the concept of length and its measurement.
2. All concepts are linked to experience through appropriate activities. Concepts are experienced before they are named. Length, for example, is introduced operationally through the procedures needed to measure it.
3. New words are introduced only as needed in order to effectively communicate on a subject. As a rule, only frequently used terms are introduced

at all. Keeping to this discipline cuts down on new vocabulary words and guarantees that new words will become familiar through frequent use.

4. Investigations progress from the more familiar to the less familiar. Investigations with the magnifying glass precede investigations with the microscope.

5. Quantitative measurements are routinely made and mathematics is used as required. For example, the investigations with the magnifying lens require measuring lengths and determining the ratio of one length to another.

Ability Grouping

The lack of focus and coherence in U.S. education damages everyone, but like all adversities, it damages the disadvantaged the most. A high-IQ student can learn from many sources, but a low-IQ student depends almost exclusively on the school. Such students don't pick up concepts from context; the difference between "and," "but," and "or" must be explicitly explained to them. Low-IQ students are undoubtedly capable of much higher levels of performance than they are currently achieving. Perhaps most can eventually pass the GED tests, but only if their education is specifically designed to achieve this goal. The goal of passing the GED tests by the end of ninth grade is explicitly a policy in the politics of intelligence (Chapter 9).

This policy requires that students be divided into at least three ability groups. All groups would have to complete the same core curriculum, but they would do so at different rates and at different levels. Level I students—those who, for whatever reason, do badly in school—would spend 80 to 90 percent of their time on the material in their basic books. They would read and analyze short passages, work on basic mathematics skills, and study science primarily as exercises in comprehension and communication. Level II students—the mainstream—would spend 50 percent of their time on basics, and the rest would be devoted to higher-level activities: longer reading and writing assignments, conducting and analyzing experiments, working on challenging mathematics problems. Level III students—the high achievers—would use the basic material mainly for review. Most of their time would be spent reading major literary works, studying mathematics and science at a theoretical level, and working on independent projects.

Although this looks like tracking, smells like tracking, and tastes like tracking, it isn't really tracking. All levels are working toward the same goal and students can move from one level to another as they, and their teachers, think advisable. Passing the GED by ninth grade will be a breeze for some students and a trial for others, and their programs of study should be adjusted accordingly. Level III students will compete with one another for top scores, while level I students will struggle to pass. Some level I students, knowing that they're not ready to pass the first time they take the tests, will set personal goals for their first efforts.

This sort of grouping by ability has long been an anathema to black leaders who fear that it will re-segregate classes (Chapter 9). Mainstreaming and inclusion, which puts students of all abilities and disabilities into the same classrooms, are the current doctrines in U.S. education. They reduce education to warehousing and will only succeed in driving the more affluent students—black and white—out of public education. In a few years, educators will reinvent ability grouping under a different name.

Already the American Federation of Teachers is demanding that disruptive and dangerous students be put into special schools and that academic standards be enforced. In many states, students will soon be required to pass a tenth-grade test, not unlike the GED, in order to graduate high school. Those who fail to pass this examination can continue to study for it, while those who do pass get a Certificate of Initial Mastery (CIM) and continue with their high school program. All that I'm suggesting is that a state, instead of developing its own test and issuing a new degree, use the GED tests to qualify students for the CIM. In this way those who pass the tests have the options of continuing in high school, going into a training program, going to work, or going to college.

It's vitally important that there be a meaningful intermediate certificate to provide young people with an honorable way to leave school after ninth or tenth grade. The drive to push everyone through twelve years of academic study has made "drop outs" of those who are unable or unwilling to do so. An intermediate certificate, backed by a test of basic eighth-grade skills, provides students and their teachers with a fixed goal that they can, like mice in a maze, learn to efficiently reach. And they will learn just as a mouse does, backwards from the goal (Chapter 8).

If students have to pass a test in ninth or tenth grade, their progress must be checked in eighth grade. Those who are behind schedule will need special classes, as will those who are way ahead; hence ability grouping. But eighth grade is rather late for serious deficiencies to be detected, so earlier tests will be given, and earlier grouping. Once there is an objective educational goal, with clear consequences for those who do and don't reach it, the entire educational system will be better focused on the needs of all students.

Education Reform

Most of the ideas expressed in this chapter are opposed by most educators. The middle-school philosophy doesn't accept that middle school is preparation for high school, let alone for life. It's purpose, in the minds of certain advocates, is to prepare students to be good middle schoolers. Such people will oppose any externally imposed goal.

On the other hand, there are teachers who believe that any student who doesn't graduate college is a failure. These teachers, mostly in the suburbs, have little knowledge of life outside the middle class. They will oppose any plan that appears to encourage students to leave school before high school graduation. They generally support high, albeit unattainable, standards.

Teachers who work in prisons or with parolees see a different world. For them it's a life or death matter to get grown men up to an eighth-grade level. Only with a GED certificate can a man hope to go to driving school and maybe someday own a rig of his own.

Education in the United States is in a state of gridlock. It's fragmented into hundreds of specialties and interest groups that compete with one another for limited resources. Like the battles between tribal peoples, this competition is very ritualized, with strict constraints on the tone and content of the struggles. Although these groups talk constantly of the need for radical reform and change, they are too enmeshed in the system to change anything. Real change must be imposed from the outside.

Education is a state-created monopoly. Taxpayers are required to support schools that students are required to attend, taught by teachers who are required to take courses taught by education professors. The diploma-granting authority, even for private schools, is conveyed by the state. The state decrees alternative certifications, such as the GED, and the state forbids those below a certain age from choosing this alternative. The system is the creation of the state legislature, and only the legislature has the power to change it. Constitutionally, Washington is out of the loop.

For many years, state legislatures have served the interests of various educational factions, becoming part of the gridlock. But in 1993, under pressure from the business community, the Massachusetts legislature, which 150 years earlier had established compulsory education and teacher training institutions, abolished the education major as a pathway to teacher certification. By legislative decree, the monopoly granted to schools of education was eliminated and the products of these schools were declared invalid. Students who wish to obtain a provisional teaching certificate must now have a liberal arts major—other than education. This tectonic change instantly transferred power from education departments to liberal arts departments. At my university, there were faculty meetings to arrange honorable surrender terms, but in the end, no prisoners were taken. The education faculty, who for decades had refused to share their monopolistic position with other faculty, lost most of its role in the training of future teachers.

The Massachusetts Education Reform Act abolished tenure and requires that teachers take continuing-education courses in order to renew their certification every five years. Continuing education has long been required of most licensed professionals, but not of teachers. This new requirement creates an enormous continuing-education market that is fostering the development of serious content courses, taught by liberal arts faculty, to compete with tiresome workshops on learning styles and cooperative learning.

Of even greater potential impact is the requirement that, before the year 2000, Massachusetts students must pass a tenth-grade examination before they can graduate from high school. At present, the only state requirement is that students take, though not necessarily pass, three years of physical education and one year of American history. Many other states are also setting examination standards, but the difficulty of each state writing its own examination is formidable. Massachusetts is spending twenty-two million dollars to develop a test based on its own vague curriculum frameworks, rather than adopting the existing GED tests.

Many educators bitterly oppose standardized testing of any type, because it doesn't assess a student's real knowledge, because it categorizes students, and because the "results obscure opportunities to honor and value individual differences" (Brooks and Brooks, 1993). But GED-like tests that ask students to interpret the meaning of paragraphs are assessing exactly the type of knowledge that students need to continue their academic education. Categorization is appropriate if it's used to assist students to obtain this knowledge. Students must pay their dues like everyone else. One can honor their individual differences only when they have acquired the minimal level of academic attainment necessary to express their thoughts in a coherent and understandable manner. That is empowerment—anything less is pandering.

The battle is joined. The business community, which must deal with the illiterate products of our K–12 system, is demanding outcome-based testing. Legislatures have responded, but the outcome is uncertain. States may back off when the first students fail their examinations, or when educators mount their counteroffensive. But, when the rhetoric gets impassioned, it's good to remember that a failed test is only an assessment that a student needs to work on his basic reading and mathematics skills before continuing with more advanced subjects.

Conclusion

Science is the connection between theory and experience, and our understanding of experience comes from our understanding of the theory. In physics, these theories are very elaborate. To understand the connection between the motion of a pendulum and the motion of a falling rock requires a theoretical exposition that takes months to teach and relies on years of prior mathematical study. Modern authors of physics textbooks follow in a tradition of theoretical exposition that dates back to Euclid and Archimedes. The genius of these writers wasn't only in the originality of their investigations, but in the care with which they arranged their theorems in logical progression.

To this day, we try to build our theories upon a minimum bases of fundamental assumptions. This is a human endeavour motivated by the way human beings think and learn. Nature itself is an unbroken whole to be sure, but we human beings have to start with a bite somewhere. There may be some arbitrariness of where somewhere is, but once started the exposition follows in a more or less natural way. In practice, there aren't many radically different starting points. SEED starts its exposition with a discussion of the measurements of length, mass, and time, just as Maxwell did a hundred years earlier (Maxwell, 1891/1954; Cromer and Zahopoulos, 1993).

This theoretical tradition is totally lacking in elementary and middle-school science textbooks, which leaves students totally unprepared for high school and college science courses. Educators, following their own naturalistic tradition, believe that students should study nature in its full complexity. They want to break down the artificial barriers between the sciences, feeling that somehow the traditional disciplinary lines interfere with students getting the "big picture."

But the division of the curriculum into physics and chemistry is no more artificial than the division of physics into mechanics and electricity. Some division is necessary for orderly exposition. The connection of the various parts to an understanding of the whole requires years of study, and is well beyond the scope of middle-school science. Most practicing science teachers realize this; the push for integrated science comes from an ideological construction unconnected to experience.[3]

This holistic mentality extends to other subjects areas as well. In art, for example, we encourage children to be creative rather than skillful, whereas in China, youngsters are taught to draw by a meticulous method that breaks every complex activity into small sequential steps. The result is that Chinese children have generalizable artistic skills that far surpass those of American children (Gardner, 1989). As a child, when I drew pictures of the brick buildings in my Chicago neighborhood, I drew the rows of bricks directly above one another. As often as I had looked at these buildings, I had to be explicitly taught to offset each row by half a brick. After studying the changing of the colors in autumn, and taking a trip to the woods, a class of youngsters was asked to draw an autumn scene. Every student colored the trees green (Brooks and Brooks, 1993).

But if the general principles of brick walls and autumn trees aren't easily inferred from experience, how much less so are the meanings of such modifiers as "even," "although," and "yet." Yet an understanding of such words is critical for attaining the eighth-grade reading level of a GED-like examination. My hope for such a test is that it will force teachers into being more explicit about the principles, rules, and theories of all their subjects—art, history, English, mathematics, and science. This is important for all students, but it's especially important for level I students. There is nothing wrong with teaching students the theory you want them to know, whether the laws of physics or of grammar, as long as they have an opportunity to connect the theory to specific experiences in and out of school.

The value of a formal education is that it provides a consistent, coherent, and universal framework of basic knowledge on which individuals can build their own understanding of the world. Attacks on claims of a universal framework have undercut confidence in the whole educational enterprise and inhibited teachers from teaching coherently. Instead, students are expected to construct their own frameworks. But these are bound to be inconsistent and idiosyncratic, leading to outlandish and dangerous interpretations of events. There can be no justification for a tax-supported public education system that doesn't teach the universal framework of historical and scientific knowledge, for without such a framework it's hard to see how a heterogeneous democratic society can resist the ever-present human impulse to split into warring clans and tribes.

Appendix

Project SEED: Science Education through Experiments and Demonstrations

Project SEED (Science Education through Experiments and Demonstrations) is a physical science enhancement program for middle-school science teachers, supported by Northeastern University, the National Science Foundation, and the Massachusetts Department of Education.[1] The core of the program is thirteen day-long, activity-based workshops on the basic concepts and principles of physical science. These begin with the basic concepts, or underpinnings (Arons, 1990), that are the foundations of physical science: length, mass, time, area, and volume (Cromer, Zahopoulos, and Silevitch, 1994). From there, they go into force, density (the ratio of mass to volume), pressure (the ratio of force to area), simple machines, motion, the earth as a planet, elements and compounds, sound, optics, temperature and heat, electricity, and electromagnetism. In all, there are over 200 activities that illustrate and develop the principles of the course. These are described in SEED's *Sourcebook of Demonstrations, Activities, and Experiments* (Cromer and Zahopoulos, 1993).

SEED starts with the simplest concepts and builds up in complexity through a logically connected sequence of concrete activities. The concreteness of the activities and their logical order ensures that there are no serious gaps in a student's understanding. This is an essential point of the SEED approach that distinguishes it from most other middle-school curricula. Any concrete activity, no matter how simple, is "real-life" and so is governed by the inexorable laws of physics. Carelessness or misunderstanding in the reading of a ruler or balance shows up in poor results, which, up to a point, students are eager to correct. The quantification of the activities is essential to this because it provides a check on the validity of the whole experiment, signaling the presence of any misunderstanding. Also, it exercises newly learned arithmetic skills and generates a sense of increasing competency.

Not all topics in science, or even in physical science, lend themselves to the SEED approach. SEED is basically pre-Newtonian science, with the addition of topics in electricity and magnetism that can be studied in the same way. SEED doesn't discuss global wind patterns, for example, because there are no concrete activities that students can do to study it. This doesn't mean that global wind patterns should never be taught, but it does suggest that such topics are less appropriate for middle-school students—especially early middle-school—than the related topics of pressure and buoyancy that they can investigate themselves.

Pressure—one of the basic concepts in the SEED program—is critical to the understanding of such diverse subjects as global wind patterns, weather, hydraulic engineering, and flying. It is a "big idea," to use constructivist terminology, because it relates many different topics. Hold your wrist up to your eyes a few seconds, and you will see the veins flatten under your skin; lower your wrist to your lap, and the veins bulge out. This phenomenon is caused by variation of the pressure in a fluid with elevation, a topic of seventh-grade science (Aldridge et al., 1994).

But the basic principle governing the pressure in fluids—Pascal's law—is subtle and nonintuitive. If holes of the same diameter are drilled at three points one centimeter from the bottom of a one-liter soda bottle, the bottle filled with water, and the cap screwed on tightly, the water will flow out equally from all the holes no matter where the bottle is squeezed. An increase in the pressure at one point in the water is transmitted equally to all other points. This is the first of a sequence of demonstrations, activities, and experiments in the SEED program that enhances the teachers' understanding of pressure, while providing them with effective teaching materials and strategies.

But what is pressure? At the elementary level, the term may be used rather loosely, but its full beauty and significance requires understanding that pressure is a force per unit area, the ratio of force divided by area (F/A). But what are force and area? An understanding of area and its measurement requires a prior understanding of length and its measurement. An understanding of a ratio as strange as F/A requires prior experience with the ratios of more elementary quantities. Thus, working back from the goal of understanding pressure, a logical curriculum virtually writes itself.

The study of pressure requires repeated use of basic mathematics. How much weight does one tire of an automobile support? (Assuming a 3000 pound car, with its weight evenly distributed over four tires, the answer is 3000 pounds divided by four, or 750 pounds.) If the pressure of the air in the tire is thirty pounds per square inch, what is the area of the tire's print on the ground? (The ground exerts a force of 750 pounds over the area A of the tire's print on the ground. But F/A is the pressure in the tire, so the area is 750 pounds divided by thirty pounds per square inch, or twenty-five square inches.) What is a rectangle of this area? (Approximately six by four inches.)

To demonstrate the power of Pascal's law and its application to hydraulics, a teacher is levitated by blowing into a few drinking straws. Using duct tape, the opening of a large heavy-duty garbage bag is tightly sealed around four drinking straws. The bag is laid flat on a table, a two-foot-by-three-foot ply-wood board is placed on top of the bag, and one participant sits crossed-legged on the board. Four other participants blow gently into the straws, slowly inflat-ing the bag and lifting the board and the first participant off the table. This dra-matically demonstrates that the small pressure exerted by blowing through the straws is transmitted equally to all points under the board and that this small pressure, multiplied by the large area of the board, produces a force equal to the weight of the person sitting on it. (The area of the board is twenty-four inches times thirty-six inches, or 864 square inches. If the person on the board weighs 170 pounds, the pressure required to lift him is 170 pounds divided by 864 square inches, or only 0.2 pounds per square inch.)

Basic to the SEED approach is the conducting of many interesting experi-ments and demonstrations that are related to the same concept. In this way, understanding of a concept develops through familiarity. By building slowly and systematically, earlier concepts, such as force and area, are available when needed to discuss a more advanced topic, such as pressure. Review of these ear-lier topics occurs within the context of the new one.

SEED is a teacher enhancement program, and not a middle-school cur-riculum. It was designed to be curriculum independent, meaning that its activ-ities could be used to supplement any physical science course. Unfortunately, current middle-school textbooks don't focus on developing an understanding of basic concepts. Some SEED-trained teachers are leaving their new textbooks in unopened cartons and having their students work entirely on SEED activi-ties. The results can be spectacular when teachers take the time to systemati-cally develop the basic concepts of length, mass, area, volume, and ratio. The key is respect for the difficulty of these ideas and appreciation of the critical role they play in all subsequent work in science.

Fortunately, the SEED activities that develop these concepts are simple, yet engaging. Middle-school children, teachers, inmates, and retired engineers are all intrigued by what can be done with a simple balance made out of drinking straws. And in back of this activity is the deep philosophical point that basic concepts in science are defined by the procedures used to measure them. Sci-ence isn't a word game; its concepts are connected to the world of experience by simple concrete manipulations.

A surprisingly large amount of science can be understood with eighth-grade mathematics: decimals, fractions, ratios, proportions, area, volume, per-cents, averages, graphs, and so on. This also happens to be the level of mathe-matics on the GED test. The goal of getting all students up to this level is challenging, but it's obtainable with focused and structured curricula.

That is what SEED is all about: focus and structure. SEED is student-cen-tered in that most of the time students are actively doing something, and it's also school-centered in that these activities have been carefully chosen to advance specific curricular objectives. Tying every concept to concrete activi-ties disciplines the curriculum, preventing the inclusion of such abstract topics as global wind patterns and atomic structure. Eventually, of course, students have to be able to study more abstract subjects. This becomes possible when, because they have done enough experiments themselves, they can understand descriptions of other people's experiments.

Notes

Chapter 1

1. The elements from hydrogen (which has one proton in the nucleus of a hydrogen atom) to californium (which has ninety-eight protons) have been discovered on earth or artificially synthesized in bulk quantities. In the case of berkelium and californium (elements 97 and 98), this "bulk" is only a microgram or so, which still contains 10^{15} (a thousand million million) atoms. Beyond this, individual atomic nuclei with ninety-nine to 111 protons have been created in heavy-ion accelerators, and nuclei with even more protons will be created in the coming years. With current technology, the fragments from a single nucleus that decays a fraction of a second after it's created can signal a new "element" (Clery, 1994). But this is an extension of the meaning of element well beyond that proposed by Lavoisier: a laboratory substance that can't be decomposed by laboratory methods. All the elements, in Lavoisier's meaning of element, are known.

2. Genes aren't actually discovered so much as located. At present, it isn't possible to locate the normal genes in human DNA, since genes are only a few percent of all DNA. Current research focuses on extensive studies of hereditary diseases that are believed to be caused by a defective, or mutated, gene. The DNA of members of families in which a disease is prevalent are studied in order to locate patterns on a chromosome that are common to members who have the disease and absent in those who don't. Dozens of researchers can spend many years tracking down a single gene. Even then, they may know only what the gene does when it goes bad, not what its normal function is.

3. There are 3×10^7 seconds in a year and in fifty million years there are $(50 \times 10^6) \times (3 \times 10^7) = 1.5 \times 10^{15}$ seconds. Assuming there are a billion monkeys on earth, they could type 10^{24} lines.

The number of different lines of grammatical English is more difficult to calculate, but it's certainly less than the number of lines with only correctly spelled English words. Assuming that there are a million (10^6) English words, and that most lines have about ten words in them, the number of grammatically correct lines is less than

$$10^6 \times 10^6 \times 10^6 \times 10^6 \times 10^6 \times 10^6 \times 10^6 \times 10^6 \times 10^6 \times 10^6 = 10^{60}.$$

Dividing the total number of possible lines (10^{110}) by the number of grammatical lines (10^{60}) gives us 10^{50}, which is the number of nongrammatical lines for every grammatical one. For all their pains, the billion monkeys have only typed 10^{24} lines in fifty million years, so it's very improbable that even one of these lines is grammatical English.

4. Hall is internationally recognized for his discovery that a transverse electric field is induced in a current-carrying conductor placed in a magnetic field. The Hall effect is commonly used today to measure magnetic fields. Some modern computerized versions of Experiment 37 in the Harvard list make use of the Hall effect to measure the distortions of springs during elastic collisions.

5. The term "Kuhnian" applies to the common interpretation of *The Structure of Scientific Revolutions*. It does not refer to Kuhn's more recent views, which are less radical (Horgan, 1996).

6. Hall's laboratory-based method of teaching physics was adopted by the National Educational Association in its 1899 report on college entrance requirements. Critics at the time complained that these "standards" were too quantitative and too focused on precise experimental methodology to meet the needs and interests of high school students (Moyer, 1976).

7. There is an old camp trick with a canoe paddle that illustrates this. In very deep water, thrust the paddle straight down into the water. Once the paddle is totally submerged, the upward force on it is greater than its weight, and it does jump out of the water. Unless, of course, the water isn't deep enough, in which case the paddle gets stuck in the mud and you have to swim or wade to shore.

8. Serious errors appear throughout the three volumes and teacher guides of *Middle School Science and Technology*. In *Patterns of Change* (BSCS, 1994a), three pages after the erroneous explanation of floating, the following garbled and totally incorrect statement occurs: "When something moving in a straight line appears to curve, but in fact it is the surface beneath it that is curving, scientists describe this phenomenon as the *Coriolis effect*." For their activity, students move a pencil in a straight line (relative to themselves) across a rotating piece of cardboard. The textbook explains: "Winds appear to us to curve because the earth is rotating beneath them." This seems to be saying that the atmosphere isn't rotating with the earth, a common misconception. The Coriolis effect arises because the atmosphere does move with the earth, but the speed of the earth's surface varies with latitude. As air moves from the equator to the poles, its equatorial speed no longer matches the speed of the earth beneath it. The Coriolis effect is an advanced topic in Newtonian mechanics and has no place in any sixth-grade science program, especially one devoted to "inquiry."

9. SEED (Science Education through Experiments and Demonstrations) is an enhancement program for middle-school science teachers. It has been supported by grants from the National Science Foundation since 1990. The principal investigators on the grants are Michael Silevitch, Christos Zahopoulos, and Alan Cromer. See the Appendix for more details.

10. This wasn't a planned study. The constructivist was invited at the request of the Massachusetts Department of Education and was permitted to present on any subject. He chose density and buoyancy because it was scheduled for the following day. During the first hour of his presentation, Zahopoulos and I put in occasional comments, but as the session got ever more confusing we saw the possibility of a comparison and purposely stopped commenting. We knew that Zahopoulos would clear up the confusion the next day, and that Leventman was scheduled to conduct an evaluation the following week, so we refrained from muddying the data any further.

The major uncontrolled variable in this study is the difference between the presenters. Zahopoulos, a Ph.D. physicist, is the director of SEED and a star teacher who has given these workshops many times, both live and on television. The constructivist was also an experienced teacher, but he doesn't have a background in physics and he didn't have much recent experience teaching the topic. His failure was predictable, if you believe that knowledge and experience count for anything in teaching. But constructivists believe that their superior methodology more than compensates for their lack of knowledge and experience. If nothing else, this study may help to disabuse others of such magical thinking.

11. There are many other uncontrolled variables in this experiment, including which end of an object a child puts into the clay and how deep he pushes. There wasn't any real discovery either. Although the children were well-behaved, only a few tested more than one or two objects. The teacher, by collecting these one or two objects from twenty children, filled two charts, but what this meant to the children, or was supposed to mean, remains a mystery.

12. Although I haven't seen a teacher speak of a bad cup, I have had a reporter ask me which were the bad radioactive isotopes.

Chapter 2

1. I shouldn't have been so surprised, since an article on the ring and cylinder problem had recently been published in the *Physics Teacher* (Dietz and Gash, 1994).

2. For a rod swinging with an average speed v, special relativity imposes a correction to the period of the order of v^2/c^2, where c is the speed of light (thirty billion centimeters per second). If the period of the rod is one second, and v is three centimeters per second, the effect of relativity is 10^{-20} second. Thus, even if an atomic clock was used to measure the period to a billionth of a second (for example, 1.001786524 s), the effect of relativity would be 0.00000000000000000001 second, which is a hundred billion times smaller than the last digit measured (0.000000004 second).

3. Phi = $(1 + \sqrt{5})/2 = 1.618 \ldots$ is one of the two solutions of the quadratic equation $x^2 - x - 1 = 0$. Starting with any two numbers, say 3 and 7, a Fibonacci sequence is obtained by making each new term equal to the sum of the last two terms. Thus, starting with 3 and 7 we get the Fibonacci sequence 3, 7, 10, 17, 27, 44, and so on. The ratio of two terms, say 44/27 (= 1.6296 ...) gets closer and closer to ϕ (= 1.618 ...) the farther one goes in the sequence (Gardner, 1994). Phi is also the ratio of line segments in some geometrical figures, because $\phi = 1 + 2\sin 18°$.

4. Matthews (1994) gives the following tenets of modest realism: 1.) Theoretical terms in a science attempt to refer to some reality; 2.) Scientific theories are confirmable; 3.) Scientific progress, in at least the mature sciences, is due to these sciences giving increasingly true descriptions of reality; 4.) The reality that science describes is largely independent of our thoughts and minds.

Chapter 3

1. The danger of high-frequency "nuclear" radiation, such as gamma radiation, is much less than one might expect, considering the fantastic uproar of journalists and politicians over reports of alleged research abuses in the 1950s. Among 8,200 Nagasaki survivors who received an average radiation dose of forty-one rad, there was no increase in leukemia cases twenty years after exposure (five cases/five expected). Among 19,000 Hiroshima survivors with comparable exposure (thirty-three rad), there were thirty-one excess leukemia cases by 1965 (thirty-nine cases/eight expected). But the Hiroshima bomb released far more neutrons than did the Nagasaki bomb, and neutrons are known to be extremely harmful. (Except for the Hiroshima bombing, human exposure to neutrons is very rare.) Thus, the difference between the Nagasaki and Hiroshima leukemia

rates may be attributable largely, if not entirely, to the difference in neutron exposure. In this case, the effect of gamma radiation at these levels of exposure is statistically invisible (National Research Council, 1972).

Furthermore, we now know that cells have considerable ability to repair damage to their DNA caused by radiation and other insults (Abelson, 1994). Thus, exposure to low levels of high-frequency radiation may be not just statistically harmless, but absolutely harmless.

In the 1950s, mentally retarded students at the Fernald School in Boston were given radioactive drugs as part of a scientific study. Although each student's accumulated dose was less than one-hundredth of the forty-one-rad dose that caused no visible harm to the Nagasaki survivors, the study received wide criticism when it was reported in 1993 (Scott, 1993). Yet from such studies comes the knowledge used in nuclear medicine today to diagnosis and treat millions of people a year (Seaborg, 1995). Campaigns to discredit all such studies gull journalists and confuse the public, resulting in policies that hamper and even eliminate normal industrial, scientific, and educational activities. I discontinued harmless nuclear experiments in the freshman physics laboratory because the regulations had become too onerous. Thus is ignorance perpetuated.

2. Low-energy photons can cause molecules to vibrate. This is harmless in itself, since the vibrational energy is soon transferred to the general random motion of the molecules as a whole. But if the process continues at great intensity, the general agitation of the molecules can become large enough to break some apart. Proteins are particularly sensitive to this process, which is called cooking. Microwave radiation cooks in this manner.

3. The magnitude of the extremely low-frequency (ELF) magnetic fields found in the exposed homes was 0.2 microtesla (μT), or 0.002 gauss. The earth's magnetic field is 200 times larger, and the field near a kitchen magnet is 10,000 times greater. Why would anyone look for danger in such a small field, when we aren't bothered by much larger ones?

The answer the investigators give is that the ELF fields are alternating sixty times a second in North America (fifty times a second in Europe). It's this rapid change in the field that is said to make ELF fields potentially harmful. Rapid compared to what? I can't tap my finger fifty times a second, but my finger isn't relevant. Radiation interacts with individual molecules, and it's the motion of the molecules that is relevant. They have many different vibrational frequencies, all millions of times greater than the fifty or sixty times a second of very low-frequency radiation. In the time a molecule makes one vibration, the ELF field hardly changes at all. To a molecule, it's just another constant magnetic field, only much smaller than the earth's.

There's no way such extremely low-frequency low-magnitude fields can have any harmful effect on living organisms, and there's no consistent epidemiological evidence that shows otherwise. For a more thorough analysis of the physics involved see Bennett (1994) and Hafemeister (1996).

4. It's interesting that present-day television picture tubes are based on a hundred-year-old technology, though it's expected that new flat-screen technology will replace the cathode-ray tube in the next fifteen years.

5. X-ray diffraction, that determined the structure of DNA, is a form of interference.

6. A lecture demonstration at Harvard University uses a rather expensive video camera to demonstrate the photon-by-photon buildup of the interference pattern (Rueckner and Titcomb, 1996).

7. Although all scientists accept the operational validity of quantum mechanics, some physicists still haven't given up on the idea of a deterministic core. These skeptics form the loyal opposition, so to speak. They provide an invaluable service by constantly searching for loopholes in the status quo. So far, all their criticisms have been answered. But I sleep better at night knowing that the loyal opposition is there, guarding me from my own complacency.

Chapter 4

1. By contrast, the Koran, which repeats many Old Testament stories, makes no attempt to place them in any chronological order. The chapters aren't even arranged in the order in which they were first spoken by Mohammed. Instead, they are arranged by length, from the longest to the shortest chapter (Koran, 1974). This imparts a timelessness to Islamic thinking, and may account for the difficulty it has had in developing a legitimate place for progress.

2. Philosophically, I'm taking the position that there are no general facts about the world that we know innately or that we can derive logically without assuming some other general facts. We surely have an innate sense of time, but this doesn't logically require us to believe that all events can be mapped onto the same time continuum. The chronology principle is a generalization from experience. It perhaps could be argued that it's the simplest generalization about time that is consistent with everyday experience, but there's no law that says one must believe in the simplest generalization or that everyday experience is an infallible guide to the truth. One could just as legitimately believe in an omnipotent god who rearranges time, and the location of my car keys, at his whim.

3. This study was done on the cebus monkey, because during evolution two parts of the primate brain—the dentate nucleus of the cerebellum and the prefrontal cortex—expanded in step with each other. This led Leiner, Leiner, and Dow (1991, 1993) to hypothesize that a bridging of these two structures may have been responsible for the development of higher cognition in primates and for speech in human beings (Blakeslee, 1994).

4. I am a descendent of the great House of Helmholtz, because I did my thesis under the supervision of Hans Bethe, who did his under Arnold Sommerfeld, who did his under Helmholtz. The Helmholtz tradition of applying physics and its methods to a broad range of problems outside the strict academic definition of physics has been preserved through the generations.

5. In the 1960s, Julesz used the computer to produce conventional two-image stereograms that required colored glasses to see as a single three-dimensional image. A simple image might be a square in front of a plane. By creating his stereograms with random dots, his images were perfectly camouflaged when viewed monocularly. But with special glasses that allowed one eye to view one of the stereograms while the second eye viewed the other, the brain merged the dots from the two stereograms into an image of a camouflaged square floating in front of a camouflaged plane. The military's interest in this is obvious. For the science of perception it demonstrated, quite unexpectedly, that the brain didn't have to recognize the image in advance in order to meld the two independent signals it received from the two eyes. That is, cognition played no role in stereovision. Interestingly, the more complex the camouflaged image was, the longer it took the brain to "see" it.

Much earlier, the nineteenth-century physicist David Brewster had noticed that wallpaper patterns sometimes appear to jump into a different plane—that is, to appear in front or in back of the wall itself. Christopher Tyler, who had worked with Julesz, combined Julesz's random dots and Brewster's wallpapering effect to generate "device-free" stereograms, the amazing autostereograms that give a full three-dimensional image when stared at long enough with both eyes (Tyler, 1994). The growing familiarity of these images should not camouflage their wonder or the depth of science and technology that has gone into creating them. They are truly a new type of vision—one that only the mind can see.

6. This number differs slightly from the number in *The Bell Curve*, probably because I missed some points in the figure in *The Bell Curve* that were covered by other points.

7. The effects of gamma radiation on health are observable only in people subject to doses many times greater than that received by the general public (See Note 1, Chapter 3). To assess the effect of low doses on people, a linear regression is used on the high-dose data. Thus, if 1 percent of a population exposed to 100 rads dies of exposure within twenty years, the analysis shows that 0.01 percent of a population exposed to one rad will die. All government standards are based on such analyses, because it's the most conservative. Almost certainly the dose-response relationship is not linear. The body has repair mechanisms that make the low-dose response less harmful than is predicted by the linear-regression analysis (Abelson, 1994).

8. In 1996, the *Boston Globe* reported that for Massachusetts workers in 1994, the median family income by educational attainment was $23,192 for nongraduates, $37,780 for high-school graduates, $48,200 for some college, $64,846 for college graduates, and $84,400 for graduate school (*Boston Globe,* 1996). This way of reporting the data emphasizes the importance of the diploma, rather than raw years of schooling, in determining income. The $10,420 increase of income with "some college" doesn't necessarily disprove the credential model, since many of the "some college" workers have two-year and three-year associate and technical diplomas.

9. Between 1973 and 1982 the snowfall in Amherst, Massachusetts and the U.S. unemployment rate had a correlation of 0.98 (Kohler, 1988). This shows that an almost perfect correlation can't prove a relationship if there's no underlying theoretical justification for the relationship. It may be surprising that ten years of Amherst snowfall data correlate so highly with unemployment, but it isn't surprising that if one looks hard enough one will find ten years of something that is highly correlated with ten years of something else. This is known as the indefinite endpoint, and is used very successfully by magicians and con artists, and is constantly confounding the unwary empiricist.

Consider a magic trick that at one point requires a member of the audience to select one of three books presented to her. If she selects the book that the magician needs selected, he of course uses it. But if she doesn't, he can ask another member of the audience to select one of the two remaining. If this second member leaves the magician holding the book he needs, he uses it. Since he hasn't announced in advance how the book is actually to be selected, the magician can continue his hocus-pocus until the book he needs appears to have been selected by the audience. This is the fallacy of the indefinite endpoint.

10. The database used in the *The Bell Curve* is the National Longitudinal Survey of Youth available on CD-ROM from the Center for Human Resource Research, Ohio State University, Columbus, OH 43210.

11. Let X be an individuals IQ in standard deviations from the average and let Y be the individuals socioeconomic status in similar units. Then in the logistic model used by

Herrnstein and Murray, the probability P of this individual exhibiting a specific negative behavior is

$$P(X, Y) = \frac{P(0, 0)e^{aX}e^{bY}}{1 + P(0, 0)(e^{aX}e^{bY} - 1)}.$$

Here $P(0, 0)$ is the probability of an average individual—one with average IQ and average SES—exhibiting the behavior. The exponential factor e^{aX} is 1 for $X = 0$ (average
IQ). If the numerical factor a is positive, e^{aX} increases for positive X (above average) and
decreases for negative X. The computer program finds the numerical values of $P(0, 0)$, a,
and b that best fit this expression to the data.

12. This device is a "poor man's" double pendulum, in that it's made without ball
bearings. The chaotic behavior of the double pendulum is well known. With ball bearings, a professionally built double pendulum will exhibit chaos for twenty seconds or
more (Cromer, Zahopoulos, and Silevitch, 1992).

13. In New-Age and postmodern literature, "linear" refers disparagingly to Western
logic, objectivity, and rationality, and "nonlinear" refers approvingly to non-Western intuition, subjectivity, and holism. I am, needless to say, using the mathematical meaning
of linear: A relation $x \rightarrow y$ is linear if $(x_1 + x_2) \rightarrow (y_1 + y_2)$.

Linear relations have a potent simplicity that leads to very rich mathematical results.
The nonlocalizing and nonchaotic behavior of quantum mechanics arise from the fact
that it is a linear theory. In general, it can be said that linear systems are always stable and
predictable, whereas nonlinear systems may be chaotic. There's a slight metaphorical
connection between the unpredictability of some mathematically nonlinear systems and
the meaning of "nonlinear" in New-Age terminology. Nevertheless, the two usages
should never be confused. New Agers are unlikely to be pleased should science ever be
able to explain irrational behavior in terms of nonlinear mathematics.

Chapter 5

1. This same pattern can be seen today in the business world. Many new businesses
are started by entrepreneurs who learned their trade working for larger companies.
Some enlightened companies deliberately encourage their ambitious younger employees to spawn new divisions, thus preventing the loss of talent and the creation of rival
enterprises.

2. Lutheran missionaries had discovered these populations in the mid-1920s, but
kept the information secret to protect the population from exploitation.

3. At the time of first contact with the Australians, the staple crop of the highlanders
wasn't the yam indigenous to the region, but the South American sweet potato. This had
presumably reached the highland through the same trade routes that brought them their
shells. Thus, in spite of their isolation, they were affected, albeit in a small way, by dis-
tant events.

4. "Kraal" is an Afrikaans word derived from the Portuguese. It has the same Latin
root as "corral," originally an enclosure for carts.

5. Besides his 1,200 concubines, Shaka had organized thousands of unmarried females into regiments that were under his military command. Thus, he personally controlled a sizeable fraction of the young women in his nation. Another large fraction
were the wives and concubines of his generals and major officials. This excess must have
had social consequences of its own. We know that Shaka became deranged after the

death of his mother, and began to execute hundreds of his concubines on suspicion of sorcery. It was this outrage, on top of many others, that motivated his assassins (Ritter, 1955).

6. This is particularly true for men in the black ghetto culture, where survival may require behaving in ways that are inimical to survival in the world outside the ghetto. Behavioral change requires some sort of migration, which may be a move "uptown," a stint in the army, or six months in the National Guard Youth Challenge Corps (Herbert, 1994).

7. Although spinning and weaving are done by machine, women still cut and sew most of the world's clothing. For millions, this employment has been the first step out of poverty.

8. Extreme hierarchy may have a negative effect on real health. A recent study shows a statistically significant increase in the age-adjusted death rate in states with the highest income inequality (Kennedy, Kawachi, and Prothrow-Stith, 1996). The mechanism for this is unknown at this time, but it isn't an effect of poverty or limited access to health care. A wealthy state like New York, with very large income inequality, has a greater death rate than a less wealthy state, like New Hampshire, with smaller income inequality. It isn't race related either. As far as is known, just having people in your state with too much money can be injurious to your health.

Chapter 6

1. The major reference for this chapter is James Bowen's three volume history of Western education (1972, 1975, 1981). Bowen is an avowed progressive, so we naturally view the conflict between progressivism and realism quite differently. It's a tribute to the breadth, depth, and objectivity of Bowen's scholarship that it can be used for purposes contrary to his own. Where I have quoted authors quoted in Bowen, I referenced both the original work and Bowen.

2. The Maya wrote on on beaten-bark paper that was coated with plaster and folded accordionlike to make books, or codices. Only four codices survived the conquest: the Dresden Codex, Madrid Codex, Paris Codex, and the Grolier Codex. The codices are concerned with religious and calendrical matters. The many cycles of the Mayan calendar provided abundant opportunities for prophecy and prognostication.

3. Objective inquiry has always been an anathema to totalitarian doctrines. In the 1930s, the Nazi physicists Philipp Lenard and Johannes Stark—both Nobel laureates—advocated an intuitive Aryan science based on race and blood rather than objectivity. Only in this way could the "truth" of National Socialism be sustained. Aryan science was a hodgepodge of contradictory ideas that mixed romantic notions of organic nature and cultural idealism with a baffling form of anti-Semitism that labeled the great advances of twentieth century physics as Jewish science. Ironically, Lenard's early work on the photoelectric effect had led the way to Einstein's photon theory of light (Chap. 3) and Stark had discovered a phenomenon—the Stark effect—that was triumphantly explained by quantum mechanics. A quarter of Germany's physicists emigrated as a consequence of Nazi policies, but most of those who remained—particularly Max Planck, Werner Heisenberg and Max von Laue—defended the professional integrity of science against political subversion. And the byzantine politics of prewar Nazi Germany foiled Stark's attempt to gain autocratic control of German science (Beyerchen, 1977).

4. Genetic material of living humans can be repeatedly collected and analyzed using ever more sophisticated techniques, so if there is a common African origin, genetic analysis will almost certainly be able to prove it. The fossil evidence, on the other hand,

is limited to what has been discovered, more or less haphazardly, so far, and whatever rare random finds are made in the future. It seems highly unlikely that a case based on limited fossils can contend much longer against a case based on unlimited genetics.

5. Boston Latin starts with seventh grade and continues through high school (grade 12). It's now part of the Boston Public Schools, open to all Boston residents through competitive examination. A school policy, now under legal challenge, sets aside 35 percent of the seventh-grade admissions for some minorities (blacks and Hispanics), but not for others (Asians). The issue is further complicated by the fact that many Boston parents send their children to private schools through grade 6 to prepare them for the Boston Latin entrance examination.

6. The same year Douglass wrote this article (1872), three men in Kalamazoo filed a law suit claiming that the law that required them to pay taxes for the public high school violated the Michigan constitution. The Michigan Supreme Court upheld the law in 1874, settling the issue of the state's right to support education through taxation (Bowen, 1981).

Chapter 7

1. Henry wrote in his journal about his 1826 visit to West Point: "One article very necasary in teaching chemestrys is found in this room viz a black board on which the student is taught the atomic theory and all algebraical formula in chemistry" (1972). Henry was a towering figure of nineteenth-century American science, making major discoveries in electromagnetism, helping with armament development during the Civil War, and serving as the founding president of the Smithsonian Institution.

2. The small market for school physics interfacing software and hardware is dominated by Vernier Software, an early competitor of EduTech. In the beginning, David Vernier focused on developing inexpensive interfacing kits aimed at the more technically competent physics teachers, whereas EduTech focused on publishing simulations, tutorials, and ready-to-use interfacing hardware and software. As the market matured, interest shifted more and more to interfacing. Vernier had the persistence and expertise to stay with this trend, adapting his software and hardware to the Macintosh and IBM PC. More recently, Vernier Software has become a leading supplier of hardware interfacing accessories for graphics calculators.

3. The issue isn't all that complex. If the position x of a moving object is known at one time t, its position x' at a later time t' is given exactly by the equation

$$x' = x + v_{av}(t' - t)$$

where v_{av} is the average speed of the body during this time interval. This equation is exact because it is just the definition of average speed.

The business of any numerical method is to estimate v_{av}. In the Euler approximation, v_{av} is approximated by v, the speed at the beginning time t. By reversing lines in her code, Aspel had approximated v_{av} by v', the speed at the later time t'. I was completely wrong in calling what she proposed absurd, since at first glance it shouldn't make any difference which of these naive approximations is used. It might even be thought better to approximate v_{av} by $(v + v')/2$, the average of the speed at the beginning and end of the interval. But of these three possibilities, only the approximation $v_{av} = v'$ is stable in that, after going from time t to t' to t'', and so on, the errors don't accumulate indefinitely in calculations of cyclical motion.

4. Today there are powerful software packages that will solve differential equations and plot the results. Should students be taught to use such packages instead of doing their own programming? There are strong arguments on both sides, though instinctively I fear that much may be lost when students skip over important steps on their educational ladder.

5. It's undoubtedly a bit generous to say that curricular changes take decades, considering the difficulty schools and textbooks have teaching buoyancy 2,300 years after Archimedes (Chapter 1).

Chapter 8

1. There is some evidence that real mice do have a small tendancey to avoid cul de sacs (MacCorquodale and Meehl, 1951).

2. If P_m is the probability of one choice at point m, then $1 - P_m$ is the probability of the alternative choice at m. If the cyber mouse makes the first choice, the model increases P_m at the end of the trial by the amount $f(1 - P_m)$, where f is a number between 0 and 1. If the cyber mouse makes the alternative choice, the model decreases P_m by the amount fP_m. Although these rules seem arbitrary, they are the simplest rules that guarantee that if f and P_m are between 0 and 1, the new probability P_m' will be between 0 and 1.

3. The reinforcement schedule Cyber 1 uses $f = 1/n$, where n is the number of steps to the goal. That is, the last decision point has $f = 1$, the second to last has $f = 1/2$, the third to last has $f = 1/3$, and so on. For example, the probabilities at the second to last point are changed from 0.5–0.5 to 0.75–0.25 or 0.25–0.75, depending on the choice made.

The reinforcement schedule Cyber 2 uses $f = 1/(2n)$, where n is the number of steps to the goal. That is, the last decision point has $f = 1/2$, the second to last has $f = 1/4$, the third to last has $f = 1/6$, and so on. For example, the probabilities at the second to last point are changed from 0.5–0.5 to 0.625–0.375 or 0.375–0.625, depending on the choice made.

4. "Understanding" is a commonly used English word which has no precise meaning. It's sometimes taken to mean the ability to apply knowledge to new situations. In this sense, it is a very high-level skill. *Benchmarks for Science Literacy* says, "Learning to solve problems in a variety of subject-matter contexts, if supplemented on occasion by explicit reflection on that experience, may result in the development of a generalized problem-solving ability that can be applied in new contexts" (American Association for the Advancement of Science, 1993). The key word here is "may." We really don't know how to help students develop a generalized problem-solving ability, or whether there is such an ability apart from mere knowledge of many different problem-solving strategies. Whatever the case, since we do know how to teach students to solve specific problems, this should be the primary focus of science education.

5. This statement from the 1994 draft of the Massachusetts Curriculum Frameworks for Mathematics doesn't appear in the much-improved 1995 draft (Massachusetts, 1995). The 1995 draft stays close to the mathematics standards developed by the National Council of Teachers of Mathematics, which maintains the intellectual integrity of mathematics.

6. What I call a "procedure" is similar in many respects to what John Dewey (1922) called a "habit" and Paul Churchland (1995) called a "prototype." "Arts," "skills," "cognitive structures" (Cromer, 1993) are other terms that are used to convey the same notion of more or less discrete and independent learned mental and motor routines.

Churchland is the most explicit in seeing all complex human (and animal) behaviour as the smooth linking together of prototypical sequences, just as language is the grammatical linking of preexisting words. In this view, complex behavior requires prior mastery of the elemental prototypes of which it's composed. One must walk before one can run.

7. I gave this problem to a small class of post–high school prison inmates. They didn't have access to windows or stair wells, and the ceiling was only 2.8 m above the floor. It took a tall black drug dealer less than a minute to solve the problem. He immediately jumped up on a lab bench, lifted the upper end of the pendulum to the ceiling, and seeing that the lower end was still on the floor, lifted the acoustic tile of the dropped ceiling and put his hand into the space above it. The other group in the class then did the same thing. Needless to say, this wasn't the solution I had anticipated. It's a beautifully concrete example of how creative problem solving requires pushing beyond conventional and self-imposed boundaries. (Another problem that requires breaking a boundary is to form four equal triangles from six toothpicks. Don't break the toothpicks!)

8. My primary objective was to coordinate the sequence of experiments conducted in the laboratory with the sequence of topics taught in the accompanying lecture course. This required that all students do the same experiment at the same time. By going from two to three students in a laboratory group, there was instantly enough equipment to do this. Moreover, the number of groups that each teaching assistant had to supervise was decreased, so each group got more attention.

Chapter 9

1. The number of nitrogen atoms in all living organisms is estimated to be about 10^{38} (Kinzig and Socolow, 1994), so the number of living organisms is much less than 10^{38}. If each organism generates a billion sex cells (gametes) a second—another vast overestimate—there would be 10^{54} gametes produced each year. Life has existed on earth for three billion years, and the earth may survive another five billion years or so. Taking ten billion (10^{10}) years for the total duration of life on earth, we get 10^{64} for our overestimate of the number of gametes that ever were and ever will be produced. But 2^{1000} equals 10^{301}, a number unimaginably larger than 10^{64}. Just as the number of lines that a typewriter can generate with fifty different keyboard characters is vastly larger than the number of lines that will ever be written (Chapter 1), so the number of possible combination of 1,000 mutated and unmutated genes is unthinkably larger than the number of combination that will ever be produced.

2. Some species of bacteria undergo a process called conjugation, in which donor bacteria ("males") transfer some of their DNA to receptor bacteria ("females"). This practice is believed to have evolved from the need to repair damage to DNA caused by ultraviolet radiation (Margulis and Sagan, 1986).

3. Most of these studies are done by examining variations in the DNA in human mitochondria, which males and females inherit exclusively from their mothers, and in Y chromosomes, which males inherit exclusively from their fathers. Variation in the noncoding DNA in mitochondria—the so-called junk DNA that has no known function—isn't subject to natural selections and so is assumed to accumulate in proportion to how old the population is. There is much less variation in the highly conserved Y chromosome. Nevertheless, independent research continues to point to a common origin of all modern human beings in Africa less than 200,000 years ago (Ayala, 1995; Gibbons, 1995; Wilford, 1995; Tishkoff et al., 1996).

4. Some of the over sixty mutated forms of the gene BRCA1 have been linked to breast cancer, but only one person in 800 in the general population carries a mutated form. The peculiar finding that one in 100 Jewish women carry one particular form

(Holden, 1995), indicates that the group has been endogamous long enough to have accumulated measurable genetic differences from the general European stock. These differences are almost always statistical in nature; one group has a little more of this mutation and a little less of that. This makes the subject of population genetics elusive, but not necessarily irrelevant to issues of social concern.

5. The binomial distribution was used to calculate how pairs of genes—normal and sickle—are distributed in an African population, because there are only three possible combinations: normal-normal, normal-sickle, sickle-sickle. In the example given earlier in this chapter, it was assumed that 20 percent of the genes were sickle and 80 percent were normal. In terms of fractions, we have $0.2 + 0.8 = 1$. The term binomial refers to the use of two fractions that add up to 1. To get the probabilities of getting various combinations of two genes (one from each parent), one writes

$$1 = (0.8 + 0.2) \times (0.8 + 0.2) = 0.8 \times 0.8 + 2(0.8 \times 0.2) + 0.2 \times 0.2.$$

The probability of getting two normal genes is $0.8 \times 0.8 = 0.64$, or 64 percent; the probability of getting one sickle-cell gene and one normal gene is $2(0.8 \times 0.2) = 0.32$, or 32 percent; and the probability of getting two sickle-cell genes is $0.2 \times 0.2 = 0.04$, or 4 percent.

On a true-false test, if the students know none of the answers, the probability of getting any one question right is 0.5, and the probability of getting it wrong is $1 - 0.5 = 0.5$. The probabilities of getting all possible combinations of right and wrong answers on a ten-question test is given by the binomial expansion of the expression

$$1 = (0.5 + 0.5)^{10}.$$

6. The normal distribution is given by the Gaussian function

$$N(x) = \frac{1}{\sigma\sqrt{(2\pi)}} e^{-x^2/(2\sigma^2)}$$

where x is the difference between a score and the average score of the population and σ (sigma) is the standard deviation (spread) of the scores. This distribution peaks at $x = 0$ (average score) and falls to zero as x increases positively (above average) or negatively (below average). This distribution applies rigorously only in cases where x is a random variable that can have any value from minus infinity to plus infinity. The use of $N(x)$ to describe distributions of test scores that aren't random and that have a narrow range of outcomes is conventional. Such a use doesn't have the natural justification that the use of $N(x)$ has for describing random processes.

7. If the students guess on seven questions, their average score will be 3.5. If they always get the remaining three questions right, their overall average will be 6.5.

8. The scores on the Armed Forces Qualifying Test that Herrnstein and Murray use for all of their analyses did not have a strict bell-shaped distribution. There is a slight pile-up of scores at the high end, because several people answered all the questions correctly (1994).

9. In Appendix IV, Herrnstein and Murray explain their discounting of r^2 this way:

> The size of r^2 tells something about the strength of the logistic relationship between the dependent variables and the set of independent variables, but it also depends on the composition of the sample, as do correlation coefficients in general. Even an in-

herently strong relationship can result in low values of r^2 if the data points are bunched in various ways, and relatively noisy relationships can result in high values if the sample includes disproportionate numbers of outliers. (1994)

I admit I didn't understand this paragraph when I read it the first time—or the second, or third time, if the truth be told. But after performing some model analyses of my own, including those in Chapters 4 and 9, it began to make sense to me. Indeed, now that I partially understand it, I'm not sure I can state it any clearer.

For example, the correlation between the live sealed-boxes test and the computerized tests was high in spite of the fact that half the students who did well on one of the tests did poorly on the other. The r^2 for the income versus education data in Chapter 4 is only 0.10—meaning that only 10 percent of the variation in income is due to education—because the sample is dominated by high-school-only workers, who have a wide distribution of income. *The Bell Curve* studied relatively rare dysfunctional behavior—such as dropping out of high school— which it found was disproportionately distributed among the relatively rare portion of the population with low IQ. In this case, tables that give the percentage of each IQ class that exhibit a particular behavior tell the story better than r^2.

As is pointed out in Chapter 4, linear regression analysis applies only to cases in which the data are, in fact, linearly related. Using arguments based on linear-regression theory to nonlinear data is bound to create error and confusion (Murray, 1995).

10. Not to leave the reader with too dismal a view of where I live, I must mention that just beyond the bagel shop is beautiful Jamaica Pond, a gem in the Emerald Necklace of Boston parks designed by Frederick Law Olmsted. And just beyond this is the two-hundred-sixty-two-acre Arnold Arboretum of Harvard University, with its unparalleled collection of 6,000 species of shrubs and trees from around the world. My office is ten minutes from my home, and Boston's theatre district is twenty minutes away.

11. There is surprising agreement on two important facts. One, blacks consistently score about one standard deviation below whites on tests of reasoning ability, though not on tests of memory or association. Two, intelligence, as measured by these tests, is inherited in whites. Do these facts prove a genetic basis for the difference in test scores between blacks and whites? If the question is asking for proof "beyond reasonable doubt," then the answer is "no." There are many reasonable people who doubt that there is a genetic basis for the test differences (Senna, 1973). Their basic argument is that we can never know all the environmental factors that may be involved, so that even studies that control for some socioeconomic factor may be missing some other small, but vital, environmental difference (Lewontin, 1973). This argument is irrefutable.

But, as we have seen in Chapter 2, one never knows all the factors in any situation. What one has, in the best of cases, is a theory that accounts for the affects of all the important known factors. In the case of race and intelligence, we have the hypothesis, or tentative theory, that there is a genetic basis for the observed difference between white and black test scores. This hypothesis is refutable; all one needs to refute it is to find the missing environmental factor that explains the difference. Since this hasn't been done, in spite of much effort, the hypothesis stands unrefuted, but not irrefutable. It can never be proved true, because no theory can be, but it will come to accepted as true by more and more scientists if the search for the missing factor continues to come up empty-handed.

For the purposes of this book, the truth or falsity of the hypothesis is irrelevant. The only important issue is that, as a practical matter, people do vary widely in a cognitive attribute needed for survival in the modern world. In largely white Massachusetts, 39 percent of all fourth graders are at, or below, the lowest proficiency level. They are just beginning to learn basic facts and procedures, but are unable to apply them to familiar situations (Lakshmanan, 1995). That the percentage is 67 percent for fourth graders in largely minority Boston doesn't make the issue of addressing the needs of these poor performers a purely racial one. Poor performers of all races need far-more-structured learning experiences than they are likely to receive under the current policy of hetero-geneous grouping.

Chapter 10

1. The statement lends itself to the interpretation that comparing a smooth texture to a rough texture is on the same level as measuring the density of an unknown sub-stance. Such a broad learning standard is no standard at all.

2. Since the last two years of high school have no measurable benefit on the lowest third of the students, and the upper third could continue their education in college after tenth grade, it is hard to see what purpose is served by grades 11 and 12. In *Uncommon Sense,* I proposed that the last two years of high school be eliminated altogether and the teachers and facilities be used to improve education in the critical middle-school years. This would better prepare the level I students for the GED tests and the level II and III students for college or technical school after tenth grade.

3. I'm not opposed to integrated science; I'm just mindful of how difficult it is to execute (Cromer, 1993). An integrated program requires that all science teachers be, in fact, science teachers. For example, they all must have microscopes in their classrooms and the knowledge and experience of how to use them with their students. Fall River has made the commitment to this laudable goal, and I'm excited to be a part of the ef-fort.

Appendix

1. The principal investigators on the grants are Michael Silevitch, Christos Zaho-poulos, and Alan Cromer.

References

Abelson, Philip H. 1994. "Risk Assessments for Low-Level Exposures." *Science* 265: 1507.

Abraham, David. 1981. *The Collapse of the Weimar Republic.* Princeton, N.J.: Princeton University Press.

Adams, Robert M. 1960. "The Origin of Cities." *Scientific American* 203(September): 153–168.

Aldridge, Bill, Russell Aiuto, Jack Ballinger, Anne Barefoot, Linda Crow, Ralph M. Feather, Jr., Albert Kaskel, Craig Kramer, Edward Ortleb, Susan Synder, and Paul W. Zitzewitz. 1993. *SCIENCE Interactions.* Columbus, Ohio: Glencoe.

American Association for the Advancement of Science. 1993. *Benchmarks for Science Literacy.* New York: Oxford University Press.

American Federation of Teachers. 1995. *A Bill of Rights and Responsibilities for Learning: Standards of Conduct, Standards for Achievement.* Washington, D.C.: American Federation of Teachers.

Archimedes. 1897. *The Works of Archimedes.* Trans. and ed. T. L. Heath. Cambridge, Eng.: Cambridge University Press. Reprinted. New York: Dover.

Armor, David J. 1995. *Forced Justice. School Desegregation and the Law.* New York: Oxford.

Aronowitz, Stanley. 1988. *Science as Power: Discourse and Ideology in Modern Society.* Minneapolis: University of Minnesota Press.

Arons, Arnold B. 1990. *A Guide to Introductory Physics Teaching.* New York: Wiley.

Arons, Arnold B. 1995. "Generalizations to Be Drawn from Results on Teaching and Learning." In *Thinking Physics for Teaching,* ed. by Carlo Bernardini, Carlo Tarsitani, and Matilde Vicentini. New York: Plenum.

Baily, Frederick G. 1977. "Steam Turbine." *Encyclopedia of Science and Technology.* New York: McGraw-Hill.

Baldwin, Janet, ed. 1995. *Who Took the GED? GED 1994 Statistical Report.* Washington, D.C.: American Council on Education.

Barasch, Seymour, et al. 1993. *GED: High School Equivalency Examination.* 13th ed. New York: Prentice Hall: Arco.

Bell, J. S. 1964. "On the Einstein Podolsky Rosen Paradox." *Physics* (N.Y.) 1: 195–200.

Bennett, Jr., William R. 1994. "*Cancer and Power Lines.*" *Physics Today* 47(April): 23–29.

Berne, Eric. 1964. *Games People Play; The Psychology of Human Relationships.* New York: Grove Press.

Beyerchen, Alan. 1977. *Scientists Under Hitler: Politics and the Physics Community in the Third Reich*. New Haven, Conn.: Yale University Press.

Blakeslee, Sandra. 1994. "Odd Disorder of Brain May Offer New Clues." *New York Times,* Aug. 2: C1.

Blakeslee, Sandra. 1994. "Theory on Human Brain Hints How Its Unique Traits Arose." *New York Times,* Nov. 8: C1.

Bohm, David. 1952. "A Suggested Interpretation of Quantum Mechanics in Terms of 'Hidden' Variables. I & II." *Physical Review* 85: 166–193.

Bohr, Neils. 1958. *Atomic Physics and Human Knowledge.* New York: John Wiley & Sons.

Bork, Alfred. 1981. *Learning with Computers.* Bedford, Mass.: Digital Press.

Boston Globe. 1996. "The Diploma Effect." January 21: 71.

Boucher, Geoff. 1994. "Beautiful Mystery." *Los Angles Times,* Aug. 26: E1.

Bovet, D., F. Bovet-Nitti, and A. Oliverio. 1969. "Genetic Aspects of Learning and Memory in Mice." *Science* 163: 139–149.

Bowen, James. 1972. *A History of Western Education.* Vol. 1. New York: St. Martin's Press.

Bowen, James. 1975. *A History of Western Education.* Vol. 2. New York: St. Martin's Press.

Bowen, James. 1981. *A History of Western Education.* Vol. 3. New York: St. Martin's Press.

Braidwood, Robert J. 1960. "The Agricultural Revolution." *Scientific American* 203(September): 131–148.

Brooks, Jacqueline Grennon, and Martin G. Brooks. 1993. *In Search of Understanding: The Case for Constructivist Classrooms.* Alexandria, Va.: Association of Supervision and Curriculum Development

Brown, Roger. 1974. "Development of the First Language in the Human Species." In *Language as a Human Problem,* ed. by Morton Bloomfield and Einar Haugen. New York: Norton.

BSCS. 1994a. *Investigating Patterns of Change. Middle School Science & Technology. Level A.* Dubuque, Iowa: Kendall/Hunt.

BSCS. 1994b. *Investigating Diversity and Change. Middle School Science & Technology. Level B.* Dubuque, Iowa: Kendall/Hunt.

BSCS. 1994c. *Investigating Systems and Change. Middle School Science & Technology. Level C.* Dubuque, Iowa: Kendall/Hunt.

Burt, Cyril. 1963. "Is Intelligence Distributed Normally?" *British Journal of Statistical Psychology* 16: 175–190. Reprinted in *The Measurement of Intelligence,* ed. by H. J. Eysenck. Baltimore: Williams & Wilkins.

Bush, George W. 1991. *America 2000. An Educational Strategy.* Washington, D.C.: Department of Education.

Caldwell, Jean. 1992. "NAACP Sues to Reform Tracking in School." *Boston Globe,* Dec. 8: 25.

Campbell, Colin. 1984. "History and Ethics: A Dispute." *New York Times,* Dec. 23: 1.

Cardon, Lon R., et al. 1994. "Quantitative Trait Locus for Reading Disability on Chromosome 6." *Science* 266: 276–279.

Carnap, Rudolf. 1966. *Philosophical Foundations of Physics.* New York: Basic Books.

Carroll, Joseph. 1996. "Pluralism, Poststructuralism, and Evoluntionary Theory." *Academic Questions* 9: 43–57.

Celis 3rd, William. 1993. "Study Says Half of Adults in U.S. Can't Read or Handle Arithmetic." *New York Times,* Sept. 9: 1.

Charleston, R. J., and L. M. Angus-Butterworth. 1957. "Glass." In *A History of Technology. Vol. III,* ed. by Charles Singer, E. J. Holmyard, A. R. Hall, and Trevor I. Williams. London: Oxford University Press.

Chomsky, Noam. 1986. *Knowledge of Language: Its Nature, Origin and Use.* New York: Praeger.

Churchland, Paul M. 1995. *The Engine of Reason, the Seat of the Soul. A Philosophical Journey into the Brain.* Cambridge, Mass.: MIT Press.

Clauser, John F. 1974. "Experiment distinction between the quantum and classical field-theoretic predictions for the photoelectric effect." *Physical Review* D9: 853–860.

Clery, Daniel. 1994. "Elements 110 is Created, But Who Spotted It First?" *Science* 266: 1479.

Cohen, I. Bernard. 1971. *Introduction to Newton's 'Principia.'* Cambridge, Eng.: Cambridge University Press.

Connolly, Bob, and Robin Anderson. 1987. *First Contact. New Guinea's Highlanders Encounter the Outside World.* New York: Penguin Books.

Crews, David. 1994. "Animal Sexuality." *Scientific American* 270(Jan.).

Cromer, Alan. 1980. *Computer-Simulated Physics Experiments.* Newton, Mass.: EduTech.

Cromer, Alan. 1981. "Stable Solutions Using the Euler Approximation." *Am. J. Phys.* 49: 455–459.

Cromer, Alan. 1993. *Uncommon Sense: The Heretical Nature of Science.* New York: Oxford University Press.

Cromer, Alan. 1994. *Experiments in Introductory Physics.* Denton, Tex.: RonJon Publishing Co.

Cromer, Alan. 1995. "Many Oscillations of a Rigid Rod." *American Journal of Physics* 63: 112–121.

Cromer, Alan, and Christos Zahopoulos. 1993. *Sourcebook of Demonstrations, Activities, and Experiments.* Denton, Tex.: RonJon Publishing.

Cromer, Alan, Christos Zahopoulos, and Michael Silevitch. 1992. "Chaos in the Corridor." *The Physics Teacher* 30: 382–383.

Cromer, Alan, Christos Zahopoulos, and Michael Silevitch. 1994. "Physical Science Fundamentals." *The Science Teacher* 61: 42–45.

Cromer, Richard F. 1991. *Language and Thought in Normal and Handicapped Children.* Oxford: Oxford University Press.

Darwin, Charles. 1859. *On the Origin of Species.* London: J. Murray.

De Jong, Marvin L. 1991. *Introduction to Computational Physics.* Reading, Mass.: Addison-Wesley.

Dembner, Alice. 1995. "Luring Top Students is Crucial Test." *Boston Globe,* March 13: Metro/Region 1.

Denman, Roy. 1995. "This Sceptered, Smug, Shortsighted Isle." *New York Times,* Jan. 18: A21.

Dewey, John. 1922. *Human Nature and Conduct. An Introduction to Social Psychology.* New York: Henry Holt.

Dietz, E. R., and P. W. Gash. 1994. "How Sharp Does a 'Knife-Edge' Have to Be?" *Physics Teacher* 32: 46–47.

Dillon, Sam. 1994. "Special Education Absorbs School Resources." *New York Times,* April 7: 1.

Dirac, P. A. M. 1958. *The Principles of Quantum Mechanics.* 4th ed. London: Oxford University Press.

Dobzhansky, Theodosius. 1973. *Genetic Diversity and Human Equality.* New York: Basic Books.

Duncan, Richard C. 1994. "World Oil, Gas and Coal Life-Cycles: Do Hubbert's 1956 Predictions Agree with 1993 Data?" 16th Annual North American Conference

United States Association for Energy Economics and International Association for Energy Economics, Dallas, Tex.

Dunn, Rita Stafford, and Kenneth Dunn. 1978. *Teaching Students Through Their Individual Learning Styles: A Practical Approach.* Reston, Va.: Reston Publishing Co.

Dunn, Rita, and Shirley A. Griggs. 1988. *Learning Styles: Quiet Revolution in American Secondary Schools.* Reston, Va.: National Association of Secondary School Principals.

Edge, R. D. 1987. *String & Sticky Tape Experiments.* College Park, Md.: American Association of Physics Teachers.

Einstein, A., B. Podolsky, and N. Rosen. 1935. "Can Quantum-Mechanical Description of Physical Reality Be Considered Complete?" *Physical Review* 47: 777–780.

Eysenck, H. J. 1971. *Race, Intelligence and Education.* London: Temple Smith.

FC. 1995. "Excerpts from Manuscript Linked to Suspect in 17-Year Series of Bombing." *New York Times,* Aug. 2:A16.

Farmelo, Graham. 1995. "The Discovery of X-rays." *Scientific American* 273(Nov.): 68–73.

Feychting, Maria, and Anders Ahlbom. 1993. "Magnetic Fields and Cancer in Children Residing Near Swedish High-Voltage Power Lines." *Am J. Epidemiology* 138: 467–480

Feynman, Richard P., Robert B. Leighton, and Matthew Sands. 1965. *The Feynman Lectures in Physics. Quantum Mechanics.* Reading, Mass.: Addison-Wesley.

Fischman, Joshua. 1996. "Evidence Mounts for Our African Origins—and Alternatives." *Sicence* 271: 1364.

Frank, Philipp. 1949. *Modern Science and Its Philosophy.* Cambridge, Mass.: Harvard University Press.

Frank, Thomas. 1995. "A Positive Way Out for Troubled Students." *Providence Journal Bulletin,* June 17: A1.

Fraser, James W. 1977. "Reform, Immigration, and Bureaucracy, 1820-1870." In *From Common School to Magnet School: Selected Essays in the History of Boston's Schools,* ed. by James W. Fraser, Henry L. Allen, and Sam Barnes. Boston: Trustees of the Public Library of the City of Boston.

Freeman, Derek. 1992. *Paradigms in Collision.* Public Lecture. Canberra: The Australian National University.

French, Howard W. 1995. "After 6 Brutual Years of War, Peace is Celebrated in Liberia." *New York Times,* Sept. 1: A1

Galilei, Galileo. 1914. *Dialogues Concerning Two New Sciences.* Trans. Henry Crew and Alfonso de Salvio. New York: Macmillan. Reprint.

Gamow, George. 1961. *One, Two, Three . . . Infinity. Facts and Speculations of Science.* Rev. ed. New York: Viking.

Gardner, Howard. 1989. *To Open Minds: Chinese Clues to the Dilemma of Contemporary Education.* New York: Basic Books.

Gardner, Martin. 1981. *Science: Good, Bad, and Bogus.* Buffalo, N.Y.: Prometheus Books

Gardner, Martin. 1994. "The Cult of the Golden Ratio." *Skeptical Inquirer* 18: 243–247.

Gergen, Kenneth J. 1985. "The Social Constructivist Movement in Modern Psychology." *American Psychologist* 40: 266–275.

Gibbons, Ann. 1995. "Out of Africa—at Last?" *Science* 267: 1272–1273.

Glanz, James. 1995. "Measurements Are the Only Reality, Say Quantum Tests." *Science* 270: 1439–1440.

Glaserfeld, Ernst von. 1992. "Constructivism Reconstructed: A Reply to Suchting." *Science and Education* 1: 379–384.

Gold, Allan K. 1989. "Judge Approves School Choice Program in Boston." *New York Times,* June 4: 33.

Goleman, Daniel. 1995. *Emotional Intelligence.* New York: Bantam.

Gooch, George P. 1913. *History and Historians in the Nineteenth Century.* London: Longmans, Green.

Gorman, Christine. 1992. "Two Swedish Studies Provide the Best Evidence so Far of a Link Between Electricity and Cancer." *Time,* Oct. 26: 70.

Gould, Harvey, and Jan Tobochnik. 1996. *An Introduction to Computer Simulation Methods. Applications to Physical Systems.* 2nd ed. Reading, Mass.: Addison-Wesley.

Gould, Stephen. 1981. *The Mismeasure of Man.* New York: Norton.

Gould, Stephen. 1994. "Curveball." *The New Yorker,* Nov. 28: Reprinted as "Mismeasure by any Measure," in *The Bell Curve Debate,* ed. by Russell Jacoby and Naomi Glauberman. 1995. New York: Random House.

Gross, Paul R., and Norman Levitt. 1994. *Higher Superstitions: The Academic Left and Its Quarrel with Science.* Baltimore: The Johns Hopkins University Press.

Hafenmeister, David. 1996. "Resource Letter BELFEF-1: Biological Effects of Low-Frequency Electromagnetic Fields." *Am. J. Physics* 64: 974–981

Halvorson, Harlyn D., and Alberto Monroy, eds. 1985. *The Origin and Evolution of Sex.* New York: A. R. Liss.

Haugen, Einar. 1974. "The Curse of Babel." In *Language as a Human Problem,* ed. by Morton Bloomfield and Einar Haugen. New York: Norton.

Heller, Patricia, Ronald Keith, and Scott Anderson. 1992. "Teaching Problem Solving Through Cooperative Grouping." *American Journal of Physics* 66: 627–644.

Herbert, Bob. 1994. "Who Will Help the Black Man?" *New York Times Magazine,* Dec. 4: 72.

Herodotus. 1954. *The Histories.* Trans. Aubrey Sélincourt. London: Penguin.

Herrnstein, Richard J., and Charles Murray. 1994. *The Bell Curve. Intelligence and Class Structure in American Life.* New York: The Free Press.

History. 1995. "The National Standards for United States History and World History." *The History Teacher* 28: 297–386.

Holden, Constance. 1995. "Jewish Breast Cancer Gene?" *Science* 269: 1819.

Holton, Gerald. 1996. *Einstien, History, and Other Passions: The Rebellion Against Science at the End of the Twentieth Century.* Reading, Mass.: Addison-Wesley.

Homer. 1946. *The Odyssey.* Trans. E. V. Rieu. Harmondsworth: Penguin House.

Horgan, John. 1996. *The End of Science.* Reading, Mass. Additon-Wesley.

Hull, C. L. 1943. *Principles of Behavior.* New York: Appleton-Century-Crofts.

Hull, C. L. 1952. *A Behavior System.* New Haven, Conn.: Yale University Press.

Hume, David. 1939. *An Enquiry Concerning Human Understanding.* Ed. Edwin A. Burtt. *The English Philosophers from Bacon to Mill.* New York: Modern Library.

Hymes, Dell. 1974. "Speech and Language: On the Origins and Foundations of Inequality Among Speakers." In *Language as a Human Problem,* ed. by Morton Bloomfield and Einar Haugen. New York: Norton.

Ingram, Mrill et al. 1993. *Bottle Biology.* Dubuque, Iowa, Kendall/Hunt.

Iona, Mario. 1994. "New Science for Middle Schools." *The Physics Teacher* 32: 44–45.

Jackson, J. B. C., and A. H. Cheetham. 1994. "Phylogeny Reconstruction and the Tempo of Speciation in Cheilostome Bryozoa." *Paleobiology* 20: 407.

Jacoby, Russell, and Naomi Glauberman, eds. 1995. *The Bell Curve Debate.* New York: Random House.

Jensen, Arthur A. 1969. "How Much Can We Boost IQ and Scholastic Achievement?" *Harvard Educational Review* 31: 1–123.

Jensen, R. V. 1992. "Bringing Order Out of Chaos." *Nature* 355: 591–592.

Jordan, Thomas F. 1994. "Quantum Mysteries Explored." *American Journal of Physics* 62: 874–880.

Julesz, Bela. 1971. *Foundations of Cyclopean Perception.* Chicago: University of Chicago Press.

Kaku, Michael. 1994. *Hyperspace.* New York: Oxford University Press.

Kennedy, B. P., I. Kawachi, and D. Prothrow-Stith. 1996. "Income Distribution and Mortality: Test of the Robin Hood Index in the United States." *British Medical Journal* 312: 1004–1008.

Kenyon, Kathleen M. 1954. "Ancient Jericho." *Scientific American* 190(April): 76–82.

Kerr, Richard A. 1995. "Did Darwin Get It All Right?" *Science* 267: 1421–1422.

Kinzig, Ann P., and Robert H. Socolow. 1994. "Human Impact on the Nitrogen Cycle." *Physics Today,* Nov.: 24–31.

Kohler, Heinz. 1988. *Essentials of Statistics.* Glenview, Ill.: Scott, Foresman.

Kolata, Gina. 1995. "Breast Cancer Gene Defect Found in 1% of Jews in U.S." *New York Times,* Sept. 29: A24.

Koran. 1974. Trans. N. J. Dawood. Harmondsworth: Penguin Books.

Krauskopf, John, and Richard Srebro. 1965. "Spectral Sensitivity of Color Mechanisms: Derivation from Fluctuations of Color Appearance." *Science* 150: 1477–1479.

Kuhn, Thomas S. 1970. *The Structure of Scientific Revolutions.* 2nd ed., enlarged. Chicago: University of Chicago Press.

Kwiat, Paul G., Klaus Mattle, Harald Weinfurter, and Anton Zeilinger. 1995. "New High-Intensity source of Polarization-Entangled Photon Pairs." *Physical Review Letters* 75: 4337–4341.

Lardner, Dionysius. 1840. *The Steam Engine Explained and Illustrated.* London: Taylor and Walton.

Lefkowitz, Mary. 1996. *Not Out of Africa.* New York: Basic Books.

Leiner, Henrietta C., Alan L. Leiner, and Robert S. Dow. 1991. "The Human Cerebro-cerebellar System: Its Computing, Cognitive, and Language Skills." *Behavioral Brain Research* 44: 113.

Leiner, Henrietta C., Alan L. Leiner, and Robert S. Dow. 1993. "Cognitive and Language Function in the Human Cerebellum." *Trends in Neuroscience* 16: 444.

Leventman, Paula. 1995. Preliminary Evaluation of 1995 SEED Program. Boston: Northeastern University Press. Unpublished.

Lewontin, Richard C. 1973. "Race and Intelligence." In *The Fallacy of I.Q.,* ed. by Carl Senna. New York: The Third Press-Joseph Okpaku Publishing Co.

Lightman, Alan 1993. *Einstein's Dreams.* New York: Warner Books.

Locke, John. 1690a/1939. *Essay Concerning Human Understanding.* In *The English Philosophers from Bacon to Mill,* ed. by E. A. Burtt. New York: Modern Library.

Locke, John. 1690b/1939. *Essay Concerning the True, Original, Extent and End of Civil Government.* In *The English Philosophers from Bacon to Mill,* ed. by E. A. Burtt. New York: Modern Library.

Lorentz, Konrad. 1966. *On Aggression.* New York: Harcourt.

MacCorquodale, K., and P. E. Meehl. 1951. "On the elimination of cul entries without obvious reinforcement." *Journal of Comparative and Physiological Psychology* 44: 367–371.

Mach, Ernst. 1942. *Science of Mechanics. A Critical and Historical Account of its Development.* 9th ed. New York: Open Court.

Machold, Dolf K. 1992. "Is Physics Worth Teaching?" *Science and Education* 1: 301–311.

Mandell, Arnold J., and Karen A. Selz. 1994. "The New Statistical Dynamics: An Informal Look at Invariant Measures of Psychological Time Series." In *Dynamical Systems in Social Psychology,* ed. by Robin R. Vallacher and Andrzej Nowak. San Diego: Academic Press.

Margulis, Lynn, and Dorion Sagan. 1986. *Origin of Sex.* New Haven, Conn.: Yale University Press.

Markowsky, George. 1992. "Misconception About the Golden Ratio." *College Mathematics Journal* 23: 2–19.

Mason, Stephen F. 1962. *A History of the Sciences.* Revised ed. New York: Collier Books.

Massachusetts Department of Education. 1994. *The Massachusetts Curriculum Frameworks for Mathematics. Review Draft.* Malden, Mass.: Massachusetts Department of Education.

Massachusetts Department of Education. 1994. *The Massachusetts Curriculum Frameworks for Science and Technology.* Malden, Mass.: Department of Education.

Massachusetts Department of Education. 1995. *Achieving Mathematical Power. Draft.* Malden, Mass.: Massachusetts Department of Education.

Massachusetts Department of Education. 1995. *Owning the Questions Through Science and Technology Education. The Massachusetts Science and Technology Curriculum Framework.* Malden, Mass.: Department of Education.

Matthews, Michael R. 1993. "Constructivism and Science Education: Some Epistemological Problems." *J. of Science Education and Technology* 2: 359–370.

Matthews, Michael R. 1994. *Science Teaching. The Role of History and Philosophy of Science.* New York: Routledge.

Matthews, Michael R. 1995. *Challenging New Zealand Science Education.* Palmerston North, New Zealand: Dunmore Press.

Maxwell, James Clerk. 1954. *A Treatise on Electricity and Magnetism.* 3rd ed. New York: Dover.

Maynard Smith, John. 1978. *The Evolution of Sex.* Cambridge, Eng.: Cambridge University Press.

McCarthy, B. 1980. *The 4MAT System.* Oakbrook, Ill.: Excel, Inc.

McDougall, Walter A. 1995. "Whose History? Whose Standards?" *Commentary,* May: 36–43.

McGervey, John D. 1995. "Hands-on Physics for Less than a Dollar per Hand." *The Physics Teacher* 33: 238–241.

Meltzer, Milton. 1965. *In Their Own Words. A History of the American Negro.* 1865–1916. New York: Thomas Y. Crowell.

Mermin, N. David. 1994. "Quantum Mysteries Refined." *American Journal of Physics* 62: 880–887.

Middleton, Frank A,. and Peter L. Strick. 1994. "Anatomical Evidence for Cerebellar and Basal Ganglia Involvement in Higher Cognitive Function." *Science* 266: 458–461.

Morris, Colleen. 1993. "Williams Syndrome: Autosomal Dominant Inheritance." *American J. of Medical Inheritance* 47: 478.

Morse, Robert A. 1992. *Teaching about Electrostatics.* College Park, Md.: American Association of Physics Teachers.

Moyer, A. 1976. "Edwin Hall and the Emergence of the Laboratory in Physics Teaching." *Physics Teacher* 14: 96–103.

Murray, Charles. 1995. "*The Bell Curve* and Its Critics." *Commentary,* May: 23–30.

National Research Council. 1972. *The Effects on Populations of Exposure to Low Levels of Ionizing Radiation.* Washington, D.C.: National Academy of Science.

National Research Council. 1992. *National Science Education Standards: A Sampler.* Washington, D.C.: National Academy of Science.

National Research Council. 1995. *National Science Education Standards.* Washington, D.C.: National Academy of Science.

National Science Teachers Association. 1993. "Second Standards Document Stresses Inquiry and Relevance." *NSTA Reports,* Feb/Mar.: 1.

Newton, Isaac. 1947. *Mathematical Principles of Natural Philosophy,* trans. by Andrew Mott; ed. by Florian Cajori. Berkeley and Los Angeles: University of California Press.

Nieves, Evelyn. 1994. "Hartford Becomes Test Case in Fighting Menace of Gangs." *New York Times,* Dec. 26: 1.

Nordenskiöld, Erik. 1928. *The History of Biology.* Trans. Leonard Bucknall Eyre. New York: Tudor.

Novick, Peter. 1988. *That Noble Dream: The "Objectivity Question" and the American Historical Profession.* Cambridge, Eng.: Cambridge University Press.

Papert, Seymour. 1980. *Mindstorms.* New York: Basic Books.

Peres, Asher. 1978. "Unperformed Experiments Have No Results." *American Journal of Physics* 46: 745–747.

Potera, Carol. 1995. "A Science Teaching System Honed in a Two-Room School." *Science* 269: 1337.

Raspberry, William. 1995. "How Education Can Begin to Get Better Grades." *Boston Globe,* Sept. 3: 69.

Rauschning, Hermann. 1940. *The Voice of Destruction.* New York: Putnam.

Raymo, Chet. 1993. "To the Know–it–alls: Baloney." *Boston Globe,* Aug. 2: 26.

Reichenbach, Hans. 1951. *The Rise of Scientific Philosophy.* Berkeley: University of California Press.

Rein, W. 1893. *Outlines of Pedagogics.* Trans. C. C. and I. J. van Liew. London: Swann Sonnenschein.

Rosen, Seymour. 1994. *Electricity and Magnetism.* Englewood Cliffs, N. J.: Globe-Fearon.

Roush, Wade. 1995. "Arguing Over Why Johnny Can't Read." *Science* 267: 1896–1898.

Rousseau, Jean Jacques. 1762/1979. *Emilé or On Education.* Trans. Allan Bloom. New York: Basic Books.

Rueckner, Wolfgang, and Paul Titcomb. 1996. "A Lecture Demonstration of Single Photon Interference." *American Journal of Physics* 6: 184–188.

Rushing, Byron. 1977. "Black Schools in White Boston, 1800–1860." In *From Common School to Magnet School: Selected Essays in the History of Boston's Schools,* ed. by James W. Fraser, Henry L. Allen, and Sam Barnes. Boston: Trustees of the Public Library of the City of Boston.

Salzsieder, John C. 1995. "Atmospheric Pressure Indicator." *The Physics Teacher* 33: 224–225.

Schele, Linda, and David Freidel. 1990. *A Forest of Kings. The Untold Story of the Ancient Maya.* New York: William Morrow.

Schmandt-Besserat, Denise. 1978. "The Earliest Precursor of Writing." *Scientific American* 238(June): 50–59.

Schweber, Silvan. 1993. "Physics, Community and the Crisis in Physical Theory." *Physics Today,* Nov.: 34–40.

Science. 1995. "Sex Determination." *Science* 269: 1822–1827.

Scott, Allen. 1993. "Radiation Used on Retarded." *Boston Globe,* Dec. 26: 1.

Senna, Carl, ed. 1973. *The Fallacy of I. Q.* New York: The Third Press–Joseph Okpaku Publishing Co.

Shavelson, Richard J., Gail P. Baxter, and Jerry Pine.1992. "Performance Assessments: Political Rhetoric and Measurement Reality." *Educational Researcher,* May: 22–27.

Shaw, Thomas A.1991. "Taiwan: Gangsters or Good Guys?" In *Deviance,* ed. by Morris Frielich, Douglas Raybeck, and Joel Savishinsky. New York: Bergin & Garvey.

Shuey, Audrey M. 1966. *The Testing of Negro Intelligence.* New York: Social Science Press.

Struewing, J. P., D. Abeliovich, and L. C. Brody. 1995. "The Carrier Frequency of the BRCA1 185delAG Mutation is Approximately 1 Percent in Ashkenazi Jewish Individuals." *Nature Genetics* 11: 198.

Sulloway, Frank J. 1995. "Birth Order and Evolutionary Psychology: A Meta-Analytic Overview." *Psychological Inquiry* 6: 75.

Sulloway, Frank. 1996. *Born to Rebel.* New York: Pantheon.

Sussman, Gerald Jay, and Jack Wisdom. 1992. "Chaotic Evolution of the Solar System." *Science* 257: 56–62.

Thurow, Lester. 1985. *The Zero-Sum Solution.* New York: Simon & Schuster.

Time Almanac. 1994. "Civil Rights: 1950s" CD-ROM. Washington, D.C.: Compact Publishing Co.

Tobias, Sheila 1993. "What Makes Science Hard? A Karplus Lecture." *Journal of Science Education and Technology* 2: 297–304.

Tyler, C. W. 1983. "Sensory Processing of Binocular Disparity." In *Vergence Eye Movements: Basic and Clinical Aspects,* ed. by C. M. Schor and K. J. Ciuffreda. London: Butterworths.

Tyler, Christopher W. 1994. "The Birth of Computer Stereograms for Unaided Stereovision." In *Stereogram.* San Francisco: Cadence Books.

Vallacher, Robin R., and Andrzej Nowak. 1994. "The Chaos in Social Psychology." In *Dynamical Systems in Social Psychology,* ed. by Robin R. Vallacher and Andrzej Nowak. San Diego: Academic Press.

Velikovsky, Immanuel. 1950. *Worlds in Collision.* New York: Macmillan.

Verhovek, Sam Howe. 1995. "A Change in Governors Stalls Model Drug Program in Texas." *New York Times,* July 4: 1.

Waldrop, M. Mitchell. 1987. "The Workings of Working Memory." *Science* 237: 1564.

Wang, Paul P., and Ursula Bellugi. 1994. "Evidence of Two Genetic Syndromes for a Dissociation between Verbal and Visual-spatial Short-term Memory." *J. of Clinical & Experimental Neuropsychology* 16: 317.

Watson, James D. 1968. *The Double Helix.* New York: Atheneum.

Wilford, John Noble. 1995. "Genetic Sleuths Follow Clues to Elusive Ancestral 'Adam.'" *New York Times,* Nov. 23: A1.

Wilkerson, Rhonda M., and Kinnard P. White. 1988. "Effects of the 4MAT System of Instruction on Students' Achievement, Retention, and Attitudes." *The Elementary School Journal* 88: 357–368.

Wilson, Edmund O. 1975. *Sociobiology.* Cambridge: Belknap Press of Harvard University Press.

Wilson, Edmund O. 1995. "Science and Ideology." *Academic Questions* 8(Summer): 73–81.

Yager, Robert E. 1991. "The Constructivist Learning Model." *Science Teacher,* Sept.: 52–57.

Yamamoto, Tomomyuki, and Kunihiko Kaneko. 1993. "Helium Atom as a Classical Three-Body Problem." *Physical Review Letters* 70: 1928–1931.

Ziman, John M. 1968. *Public Knowledge.* Cambridge, Eng.: Cambridge University Press.

Index